Lecture Notes in Physics

165

N. Mukunda H. van Dam
L.C. Biedenharn

Relativistic Models of Extended Hadrons Obeying a Mass-Spin Trajectory Constraint

Lectures in Mathematical Physics at
the University of Texas at Austin
Edited by A. Böhm and J.D. Dollard

Springer-Verlag
Berlin Heidelberg New York 1982

Authors
L.C. Biedenharn
Physics Department, Duke University
Durham, NC 27706, USA

H. van Dam
Physics Department, University of North Carolina
Chapel Hill, NC 27514, USA

N. Mukunda
Centre for Theoretical Studies, Indian Institute of Science
Bangalore-560012, India

Editors
A. Böhm
Physics Department, University of Texas
Austin, TX 78712, USA

J.D. Dollard
Mathematics Department, University of Texas
Austin, TX 78712, USA

ISBN 0-540-11586-2 Springer-Verlag Berlin Heidelberg New York
ISBN 0-387-11586-2 Springer-Verlag New York Heidelberg Berlin

Printing and binding: Beltz Offsetdruck, Hemsbach/Bergstr.
2153/3140-543210

Preface

The purpose of the Texas lecture notes is to inform graduate students and "non-specialists" about recent developments in various areas of mathematics and physics. This volume of notes originated from two series of lectures which were presented by two of the authors during the academic year 1979-80 at the University of Texas at Austin.

The subject discussed in this volume is one of the most fundamental problems of particle physics - the problem of the hadron mass spectrum. It is the question of how the continuous parameter mass depends upon the discrete parameter spin, resulting in the discrete mass spectrum for the hadrons.

There is no generally accepted solution to this problem nor is there a generally accepted proposal on how to attack it. The proposal discussed here uses the methods of representation theory. It is based upon the idea that the internal structure of the hadrons is not explicitely given in terms of elementary constituents but is more abstract.

Due to the novelty of this approach, the description of a relativistic quantum mechanics for extended objects proposed in these notes (in particular Chp. 4), may not be in its final form. Still, it warrants publication as a review since it will continue to stimulate discussions and further developments in this important area.

Chapters 5 through 8 are mainly concerned with the application of constrained Hamiltonian mechanics and give a very clear and beautiful presentation of classical models for relativistic rotating objects. If one believes that quantum theory is obtained by the correspondence principle from classical mechanics, this is a subject that cannot be avoided.

The Editors

CONTENTS

CHAPTER ONE

INTRODUCTION

One of the significant early developments in hadron spectroscopy was the empirical realization that excited hadronic states--having the same SU3 labels and parity--could be organized into families called "Regge sequence" or "Regge trajectories". The resulting $(\text{Mass})^2$ versus spin plots, developed by Chew and and Frautschi, were expected (from potential model calculations) to show strong curvature, but as the data were assembled straight line fits, $M^2 = C_o s$, with a universal slope ($C_o \sim .1\ \text{GeV}^2$) seemed indicated. Such linearly rising Regge trajectories became an acceptable empirical model for hadronic structure, which could be considered to agree with unitarity constraints as a sort of 'zero-width first approximation' [MAN 1]. Coupled with the concept of duality, this viewpoint--as is well known--led first to the Veneziano model, then to dual resonance models, and eventually to the technically very impressive dual string models [FRA 1].

An alternative line of development of the trajectory concept came from the discovery of SU(6) symmetry [DYS 1], combining Gell-Mann SU(3) with the intrisinic spin SU2, in a way familiar from Wigner's introduction of (*non-relativistic*) SU4 symmetry in nuclear structure physics. In hadronic physics, SU6 symmetry is problematic, since it appeared justifiable only non-relativistically, but was applied in a relativistic context. More exotic models were developed in great profusion, most--unfortunately--wholly incorrect (such as $\tilde{U}(12)$) and in obvious contradiction to basic principles.*

This line of development was brought to a halt by the McGlinn-O'Raifeartaigh theorem which asserts (in essence) that: If one has a finitely generated Lie symmetry group G which contains the Poincaré group P as a subgroup (and G is not a direct product G' x P), *then* in any UIR of G either the mass spectrum is continuous, or consists of a single discrete point. Any hope of producing the empirical Regge trajectories via the representations of a symmetry group seemed doomed.

The reaction to this theorem was almost as extreme as the $\tilde{U}(12)$ fiasco itself, which the theorem had eliminated. This was especially unfortunate, since the use of group structure as a substitute for dynamics is a valid--and very useful--theoretical technique for model building, and moreover, some of the symmetry models being developed

---

*S. Coleman, Proc. Midwest Conference on Theoretical Physics, Indiana University, Bloomington, Ind., 1966.

at this time [NAM 1][TAK 1][BOH 1] were actually valid ways to circumvent this "no-go' theorem. Nevertheless, the belief is still widespread that only the use of supersymmetry (boson-fermion multiplets) avoids the no-go theorem.

The models we shall discuss in detail in these lectures follow the line of development of symmetry structures; these models were clearly foreshadowed, more or less explicitly, in a brief survey by Takabayashi in 1967 [TAK 1]. These same models may also be considered in terms of the other line of development since--in the limit--they fit into string model structures.

In order to treat, relativistically, structures which are *not* elementary--as for example an entire Regge trajectory--, one is forced to make a choice right at the beginning: (a) is the structure to be made up of other (possibly elementary) particles? or (b) is the "internal structure" to be more abstract (for example not realizable as particles?) If one chooses (a) then one faces serious, as yet not fully solved, problems: cluster decomposition, existence of a continuum,... and essentially the only way to proceed is via a fully fledged field theory. (Cf. however, the recent work of Komar [KOM 1], F. Rohrlich [ROH 1] and I. T. Todorov[TOD 1].)

The fact that quarks are not (as yet) seen, leads one to hope that alternative (b) is a valid approach, for which, in particular, there is no continuum and the indefinitely rising discrete Regge trajectory is an acceptable first approximation.

The essential point in this use of alternative (b) is that the internal coordinates, while varying under Lorentz transformations (and hence carrying spin angular momentum) do not carry *any linear momentum.** (This is not quite the same as removing the center-of-mass coordinate, having translation invariant internal coordinates, since (by reversing the process) one could recover the linear momentum carried by the separate constituents.) The internal (spin) angular momentum is defined by spin operators acting on the abstract internal coordinates and these *do not* arise from differences of particle positions. (This is discussed in Chapter 2.)

Such a viewpoint is a literal generalization of the concepts used in the four-component Dirac wave function. There is one, and only one, position coordinate (x_μ) and the spin is carried by the indices. Generalizing to a denumerable infinity of indices is equivalent to using:

---

* This circumstance is less surprising if one recalls that phonons (of non-zero frequency) also do *not* carry linear momentum, as Peierls has discussed [PEI 1]. This analogy would suggest that the internal variables used here are Fourier modes of the internal structure.

$\psi_n(x_\mu) \to \psi(x_\mu;\xi)$, where we have introduced in ξ one (or more) continuous internal variables. (Hence the term: infinite component wave functions.)

Such a generalization was suggested very early (]950's) by Yukawa [YUK 1] as bi-local (later quadri-local) fields. Yukawa used as internal variables *four-vectors*, and this leads at once to an insuperable problem for interactions. (Constraints are required to remove redundant timelike variables and the associated negative energy timelike oscillations, and interactions are incompatible with the constraints, see Chapters 5 and 6.)

The proper way to proceed--or at least a way which is successful in allowing interactions--is to use *spinorial* internal variables. This is, once again, an idea that appears in the early literature--nost notably in the Majorana equation [MAJ 1],[FRD 1],[GEL 1]--but the *bad spectral properties* (accumulation points in the mass spectrum at zero-energy,...) led to an equally early discard. For another approach see [DRE 1].

Our interest in this already well-studied field was stimulated by a paper by Dirac in 1970 [DIR 1]. In this paper, Dirac gave a new wave equation with spinor internal variables that *superficially* was similar in appearance to the famous Dirac electron equation. The equation was, in fact, a very ingenious way to avoid the spacelike (and light-like) solutions of the Majorana spectrum, and it had an inherently positive energy spectrum, a conserved current, but *forbade any (minimal) coupling to electromagnetism.* An especially intriguing point was that Dirac claimed that in a moving system the "particle" appeared to have *all* spins. This is an error*, resolved later by Dirac himself [DIR 2] and is due to a misidentification of the spin. Alternatively, one may say that the orbital angular momentum is misidentified by using the *wrong position coordinate.*

This latter is in itself a very interesting idea, for it suggests that we are dealing in Dirac's new model with *two* positions: the position of the "*charge*" (the Minkowski position, x_μ) as contrasted with the position of the *center-of-mass* (the coordinate defining, through the orbital angular momentum, the proper spin). These ideas are physically appealing and quite suggestive, for they are just the old Zitterbewegung concepts introduced by Schrödinger [SCH 1], [DIR 3], but now realized in the context of a (positive energy) extended structure. (We discuss this in detail in Chapter 4 and, more physically, in Chapter 8.)

We may summarize now the essential ideas on which the models of these lectures are based and then outline the contents of the successive chapters:

* It is curious to note that in Dirac's 1949 paper on expansors the same (erroneous) claim appears [DIR 4].

(a) a discrete, rising, unbounded mass spectrum (versus spin) is taken for the Regge trajectory constraint;

(b) internal variables are abstract (unobservable), translation-independent, *spinorial* coordinates;

(c) the internal variables are oriented by the state of motion (Lorentz frame) of the object (see Chapter 2, §4.);

(d) the structure is a generalization of Dirac's new equation to encompass all spins. (Dirac's structure had only spin 0.)

The fact that center-of-mass and charge positions are not the same (in the models we treat) introduces another basic theme of these lectures: the relation between the quasi-Newtonian (or Newton-Wigner) [NEW 1] coordinates and the Minkowski position coordinates. The general concept stems from the Foldy-Wouthuysen transformation for the Dirac electron equation [FOL 1], [PRY 1], [BEC 1]: the center-of-mass coordinates are the "mean position"coordinates, the related orbital angular momentum and spin, the "mean orbital angular momentum" and "mean spin" respectively. These latter angular momenta are each separately conserved. Although fully relativistic, these quasi-Newtonian position coordinates are not *manifestly* relativistic; in the particular they do not constitute the space components of some four-vector. Moreover, *nothing couples to the quasi-Newtonian coordinates.* The troubles of "relativistic SU(6)" are, in fact, precisely the troubles of quasi-Newtonian coordinates. We discuss in Chapter 3 the general subject of quasi-Newtonian coordinates, developing these ideas from the Wigner form of the Poincare irreps for massive particles with arbitrary spin [WIG 1].

The problem of transforming from quasi-Newtonian to Minkowski coordinates is discussed in Chapter 4. Here we develop the "inverse Foldy-Wouthuysen" transformation, but in our case for all spins at once. The successful, explicit, realization of this "inverse F-W" transformation (for the entire set of (M(s),s) Wigner irreps) should be an instructive example for more complicated--but closely related--situations (such as the Melosh transformation between "current quarks" and "constituent quarks" [MEL 1].

The end result achieved in Chapter 4 is a constrained Lagrangian for an extended object in Minkowski space. To study this structure in the required detail we develop in Chapter 5, the powerful techniques of Dirac for analyzing constrained Lagrangian-Hamiltonian mechanics. In this, our work again has predecessors: in this case the Hanson-Regge [HAN 1], [HAN 2] treatment of the relativistic top. Their discussion is based on internal coordinates built from vectors (a Lorentz tensor) and--unlike the spinor model we develop--fails in both interactions (constraints) and in the use of non-commuting Minkowski coordinates for the structure to be quantized. (The Hanson-Regge model is dicussed in Appendix A.) Using the techniques of Chapter 5 we discuss (in Chapter 6) vectorial models and (in Chapter 7) the spinorial model, which--as we show--permits of electromagnetic interactions. In this chapter, we show that minimal electromagnetic coupling results in *anomalous magnetic moments*.

In the concluding chapter, Chapter 8, we exploit the fact that the spinorial model has a well-defined classical structure (classical relativistic mechanics) to develop a clearer physical picture underlying the free particle motion. For example, it is by no means clear, a priori, that (unobservable) abstract coordinates, not realized in Minkowski space, can give the particle "spatial extension" *in Minkowski space*. Just this is, in fact, the case: for it is the spin motion that causes the charge position to differ from the center-of-mass position and this separation--the spatial extension--is proportional to the spin (and hence via the trajectory to the mass). In this structure spin is literally a type of 'orbital' motion and this fact underlies the existence of an anomalous magnetic moment.

Equations of motion describing a *classical relativistic particle* with charge and with a magnetic moment in interaction with an external field--such as the structure discussed in Chapter 8--have been the subject of very many papers in the literature. There were early arguments by Bohr which cast doubt on the validity of such a classical description of spinning particles [MOT 1]. In particular, Bohr showed that such a description can only be correct if the magnetic moment of the particle is large compared to eh/mc. Similarly it was argued that the Stern-Gerlach effect cannot be observed for electrons as distinct from atoms. That such equations may be quite useful was demonstrated by the experimental use of the (Thomas)-Bargmann-Michel-Telegdi ("BMT") equation [BAR 1] in determining the g-factor of electron and muon and has been defended further [DIX 1].

The B.M.T. equations are incomplete in that they neglect derivatives of the electromagnetic field and do not include the Stern-Gerlach effect. The attempt by Good [GOO 1] to complete these equations to include field gradients was unsuccessful as his equations conflict with the conservation laws. Suttorp and de Groot [SUT 1] gave the correct equations. Van Dam and Ruygrok [VAN 1] recently derived the correct equations using the source terms by which the particles act back on the field and using the 10 conservation laws of linear and angular momentum for field plus particle.

Historically Frenkel [FRE 1] was the first one to attempt to write relativistic equations for spinning particles in an electromagnetic field. Frenkel's equations are consistent with the conservation laws, as follows from the fact that they may be derived from a Lagrangian. However, these equations are not satisfactory in that they contain third derivatives of the position. (This is due to the fact that one demands $\sigma_{\mu\nu}\dot{x}^{\nu} = 0$ instead of $\sigma_{\mu\nu}p^{\nu} = 0$, where $\sigma_{\mu\nu}$ is the spin.) These third derivatives lead to a helical motion for a free particle [WEY 1] (somewhat reminiscent of our spinor model as discussed in Chapter 8.) The radius of the helical motion could always be taken small, initially, and weak fields would not change it much. However, a transit through a strong field would sometimes give a free particle with a large helical motion. (Equations of the Frenkel type have been discussed by many authors [COR 1].)

The classical relativistic motion of our extended object, as discussed in Chapter 8, is we feel more satisfactory than any of these earlier models. Not only is the B.M.T. equation reproduced automatically, but the correct results for higher field derivatives--such as the Stern-Gerlach effect--are also obtained. The fact that these classical results are so satisfactorily obtained from the structure encourages us to the usefulness of the spinor model quantum mechanically.

Let us now compare and contrast the structures developed in these lectures with similar structures in the literature.

There is first of all a remarkable parallel between the relativistic wave equation for the spinor model and Dirac's electron equation. This is shown in the following Table 1.

Table 1

	Spinor Model:	Dirac Electron Equation:
Spin, s	$s = 0, 1/2, 1, 3/2, \ldots$	$s = 1/2$
Mass, m	$m = m(s)$ Regge	m
Minimal Electromagnetic interaction	yes	yes
Gyromagnetic ratio	$g = \frac{\partial \ln m}{\partial \ln s}$	$g = 2$
Negative energy states	no	yes
"Zitterbewegung"	"rotational"	oscillates between positive and negative energy
Classical Action	yes	no
Classical Limit	yes	not so clear

Our structure is based on a degenerate pair of harmonic oscillators which are used to give the particle internal structure. This basis is Dirac's remarkable representation of SO(3,2) [DIR 5], [MAJ 1].

Our construction can be started either quantum mechanically (Ch.3,4) or classically (5,6,7,8).

(i). *Classical construction.* This starts properly from a classical Lagrangian, which is, however, singular. The construction is rather analogous to the attempt by Regge and Hanson to give a relativistic description of a spinning particle using internal variables. The model of Regge and Hanson is discussed in Appendix A. The difference between our model in Chapter 7 and the model of Regge and Hanson is that we use an internal spinor variable, whereas they use an internal antisymmetric tensor. Another model using an internal vector is discussed in Chapter 6. The motivations behind setting up our vector model are twofold: we want to show how singular Lagrangian mechanics can lead to Regge trajectories quite directly, and also to exhibit the problems of Hanson and Regge in a much simpler context. The properties of these models are summarized in Table 2.

Table 2

	Spinor Model	Vector Model	Tensor Model
Minimal Electromagnetic interaction (classical)	yes	yes	yes
Produces gyromagnetic ratios	yes	yes	no
Regge Sequence	yes	yes	yes
Center of Mass differs from center of charge	yes	yes	no
Dirac Bracket $\{X_\mu(s), X_\nu(s)\} = 0$ (s is proper time)	yes	no	no
Quantization in external electromagnetic field	yes	blocked because of preceding	blocked because of preceding

It is important to note that the classical spinor model has a circular motion of the charge for nonzero spin. (This motion is discussed in detail in Chapter 8.) This circular motion creates a magnetic moment for all the states and also leads to radiation. Quantum mechanically one would expect the spin zero and spin one-half states to be stable, however. The circular motion in space is a helical motion in space time. This kind of motion had been obtained by Dirac [DIR 2] and by Staunton (and Browne) [STA 1], [STA 2] in connection with equations of motion for the spin zero state and for the spin 1/2 state of our system respectively. This elegant work is incomplete, however; one uses two constraints which together determine the level. The Hamiltonian is then arbitrary in that, in principle, it involves both constraints. This indeterminacy makes the classical motion undetermined, as it depends on the choice of Hamiltonian. As we, however, have a Lagrangian we obtain a unique classical limit with well-defined evolution instead.

(ii). *Quantum Mechanical Construction.* This is done in Chapters 3 and 4 for the free equation. The pair of degenerate oscillators,--which comprise the internal structure--may be in any of their excited states. Let the total number of quanta in the (two-dimensional) oscillator be n, then the spin is given by $s = n/2$ with the $2s + 1 = n + 1$ states of the same n giving the various orientations of the spin. The mass is given by a function of the total number of quanta. This function is essentially free and can be used to fit the Regge sequence.

In this model there are various things to be remarked:

(a) an immediate connection to supersymmetry, (Chapter 4);

(b) circumvention of O'Raifearthaigh's theorem: relativistic SU(6) is possible by taking three pairs of degenerate harmonic oscillators, (Chapter 4);

(c) the spectrum of the string model can be obtained easily by taking an infinite set of pairs of harmonic oscillators;

(d) the Majorana equation [MAJ 1] is *not* contained in our scheme; this is discussed under exceptional cases in Chapter 7.

Appendix B contains the form of our quantum mechanical model in the so-called 'front form' (light plane coordinates). In the front form 8 of the 10 generators of the Poincaré group are given kinematically, only 2 are determined dynamically. A nice physical picture of a composite particle in a Galilean subworld emerges inside this front.

Although we have constructed a classical Lagrangian theory and a relativistic quantum theory of a single particle with all spin states--which can interact with an external electromagnetic field--we lack a quantum field theory of this object. The problem is not to construct a free field theory (which is fairly straight-forward), but is rather the problem of constructing interactions.

Although we have not attempted to do so in these lectures, we believe the spinor model would be a useful model for discussing hadron phenomenology. This model (in its relativistic SU(6) version) automatically incorporates the use of the "velocity operator", P/M_{op}, in treating--with the correct kinematics--the effects of SU(3) mass splittings on weak and electromagnetic hadronic transitions. The use of velocity (rather than momentum) is a typical feature of the kinematic symmetry group approach exemplified in the spinor model;

let us remark that this feature was advocated early by Werle [WER 1]. Bohm and Teese [BOH 2] have applied similar ideas in their survey of weak interaction data where large mass splittings occur.

The material incorporated in these lecture notes was presented in seminars in the Department of Physics, University of Texas, Austin, Texas in May 1978 and in a second series, May 1980. We are grateful to Professor Arno Bohm for organizing the Mathematical Physics seminars, to members of the Physics Department for their helpfulness, and to our auditors for critical questioning. We wish to thank Professor W. Beiglböck, University of Heidelberg, (Managing Editor, Lecture Series, Springer-Verlag) for his interest and courtesy.

CHAPTER TWO

DIRAC'S NEW RELATIVISTIC WAVE EQUATION AND ITS GENERALIZATION

The original motivation for the work described in these lectures came from a new, positive energy, relativistic wave equation presented by Dirac in 1971 [DIR 1,2]. It is therefore appropriate that we begin with an account of Dirac's work. We present this in some detail for several reasons: (i) the new equation is quite elegant and of intrinsic interest (ii) it gives us a chance to set up a notation we shall use later; (iii) there is a superficial resemblance between the usual relativistic wave equation for the electron and the new equation, and it is important that one see clearly that they describe very different things. The new equation suffers, however, from a serious defect: *interaction with the electromagnetic field via minimal coupling is not permitted.* One of our goals will be to set up a theory in which this problem does not arise.

We will begin by first presenting Dirac's new equation in just the distinctive way in which he originally discussed it. Following that we will digress to discuss the technical background [the group structure of SO(3,2)] required for a deeper analysis of his equation, and then, in Section 3, to generalize the structure.

§1. DIRAC'S PRESENTATION OF HIS NEW EQUATION

Let us assume that the internal degrees of freedom involve two harmonic oscillators. The dynamical variables describing these oscillators will be denoted by the (dimensionless*) hermitian canonical variables (ξ_1, π_1) and (ξ_2, π_2); for convenience these four dynamical variables are collectively denoted by $\{Q_a\}$, a = 1,2,3,4 with $Q_1 \equiv \xi_1$, $Q_2 \equiv \xi_2$, $Q_3 \equiv \pi_1$, and $Q_4 \equiv \pi_2$.

The dynamical variables for the two oscillators obey the canonical Heisenberg commutation relations, that is,

$$[\xi_j, \xi_k] = [\pi_j, \pi_k] = 0, \quad [\xi_j, \pi_k] = i\delta jk; \quad j,k = 1,2 \tag{2.1.1}$$

Expressing these relations in terms of the variables Q_a we find that:

$$[Q_a, Q_b] = i\,\beta_{ab}; \quad a,b = 1,2,3,4. \tag{2.1.2}$$

where the 4 x 4 matrix (β_{ab}) has the form:

---

* The dimensionless variables are explicitly $\xi = (m\omega\,\hbar)^{\frac{1}{2}}x$ and $\pi = (m\omega\,\hbar)^{-\frac{1}{2}}p$ for the oscillator whose Hamiltonian is $H = p^2/2m + m\omega^2 x^2/2$.

$$\beta = \begin{pmatrix} 0 & 0 & 1 & 0 \\ 0 & 0 & 0 & 1 \\ -1 & 0 & 0 & 0 \\ 0 & -1 & 0 & 0 \end{pmatrix} \tag{2.1.3}$$

Note that β is real, skew-symmetric and obeys $\beta^2 = -\mathbb{1}$.

Dirac now states his new positive energy relativistic wave equation in a form designed to be similar to the usual electron equation. Writing the variables $\{Q_a\}$ as a column matrix Q, the new equation reads:

$$\left(\frac{\partial}{\partial x_0} + \sum_{r=1}^{3} \alpha_r \frac{\partial}{\partial x_r} + \frac{mc}{\hbar} \beta \right) Q\Psi = 0. \tag{2.1.4}$$

In this equation $x_0 = ct$ (with x_1, x_2, x_3 being the usual spatial variables) and the α_r are three 4 x 4 matrices that anti-commute with each other and with β. The α_r are also to obey $\alpha_r^2 = \mathbb{1}$. The mass of the particle is m.

The wave function Ψ is a function of the space-time variables (x_0, x_r) as well as a function of any two (commuting) internal variables, for example, Ω_1 and Ω_2.

This new wave equation, eq. (2.4), strongly resembles the Dirac electron equation. (In fact, if the column vector $Q\Psi$ is for a moment replaced by four *independent* functions of x instead of $Q_a\psi$ as in (2.4) then the equation becomes exactly the usual Dirac equation.) However, the the equation has only one component, ψ, with two internal variables, and thus is actually very different from the usual equation. (As we will see shortly, the spin of the new system is in fact zero.)

The four matrices (β, α_r) are an unusual realization of the Dirac matrices. There are in fact extra conditions on the $\{\alpha_r\}$ required for consistency: namely, the α_r are to be real and symmetric. There are many suitable realizations, but Dirac chooses the system:

$$\alpha_1 = \begin{pmatrix} 0 & 0 & -1 & 0 \\ 0 & 0 & 0 & 1 \\ -1 & 0 & 0 & 0 \\ 0 & 1 & 0 & 0 \end{pmatrix}, \quad \alpha_2 = \begin{pmatrix} 0 & 0 & 0 & 1 \\ 0 & 0 & 1 & 0 \\ 0 & 1 & 0 & 0 \\ 1 & 0 & 0 & 0 \end{pmatrix} \text{ and } \alpha_3 = \begin{pmatrix} 1 & 0 & 0 & 0 \\ 0 & 1 & 0 & 0 \\ 0 & 0 & -1 & 0 \\ 0 & 0 & 0 & -1 \end{pmatrix} \tag{2.1.5}$$

Consistency of the New Equation. If we adjoin the matrix $\alpha_0 \equiv \mathbb{1}$ then eq. (2.1.4) may be written in the concise form:

$$(\alpha_\mu \partial^\mu + (\tfrac{mc}{\hbar})\beta)\; Q\Psi = 0, \tag{2.1.6}$$

where $\partial^{\mu} = \frac{\partial}{\partial x_{\mu}}$, $\mu = 0,1,2,3$.

This is actually a set of four separate equations, as can be seen clearly if we define:

$$T_a \equiv (\alpha_{\mu}\partial^{\mu} + (\frac{mc}{\hbar})\beta)_{ab}\Omega_b. \qquad (2.1.7)$$

Then eq. (2.1.6) becomes:

$$T_a \Psi = 0; \quad a = 1,2,3,4. \qquad (2.1.8)$$

This makes it very clear that we have four separate equations obeyed by the single function Ψ.* Consistency of such a system of equations implies that ψ must also obey

$$[T_a, T_b]\Psi = 0, \quad a, b = 1,2,3,4. \qquad (2.1.9)$$

This commutator is easily evaluated using eq. (2.1.2) and the properties of the α, β matrices. One finds that:

$$[T_a, T_b] = i\hbar\beta[\partial_o^{\,2} - \vec{\partial}^2 + (\frac{mc}{\hbar})^2], \qquad (2.1.10)$$

so that eq. (2.1.8) implies:

$$[\partial_{\mu}\partial^{\mu} + (\frac{mc}{\hbar})^2]\Psi = 0. \qquad (2.1.11)$$

Thus *if* Ψ *obeys the new Dirac equation,* (eq. (2.1.4)), *then* Ψ *necessarily obeys the Klein-Gordon equation for mass m.* Accordingly the new equation describes free particles with mass m, but the question as to spin (and sign of the energy) is yet to be determined.

An Eigenstate of Momentum and Energy. We seek now a solution of eq. (2.1.4) corresponding to sharp four-momentum. Denoting this momentum by p^{μ}, the wave function Ψ takes the form:

$$\Psi(x;\ \xi_1\xi_2) = e^{ip\cdot x/\hbar}u\,(p;\xi_1\xi_2). \qquad (2.1.12)$$

From eq. (2.1.7) we see that $p_{\mu}p^{\mu} = m^2$; the four-momentum p^{μ} is therefore a (numerical) time-like vector, and we may by a Lorentz transformation go to the rest frame where $p^{\mu} \rightarrow \underset{o}{p}^{\mu} = (0\ 0\ 0\ \pm mc)$.

---

* Actually only three of the equations are independent since $\Omega_2T_1 - \Omega_1T_2 = \Omega_4T_3 - \Omega_3T_4$.

For this case, the four operators T_a of eq. (2.1.7) reduce (for either sign of the energy) to only two independent operators, namely:

$$(\xi_1 \pm i\pi_1)\ u(\underset{\circ}{p};\ \xi_1,\xi_2) = 0, \qquad (2.1.13a)$$

$$(\xi_2 \pm i\pi_2)\ u(\underset{\circ}{p};\ \xi_1,\xi_2) = 0 \qquad (2.1.13b)$$

where $\underset{\circ}{p}^0 = \pm$ mc.

The combinations $(\xi \pm i\pi)$ are just the "boson" operators usually denoted by a and a^+ with $[a,a^+] = 1$, $[a,a] = [a^+,a^+] = 0$.

For the case where the energy is *positive*, eqs. (2.1.13) assert that:

$$a_i u(p;\ \xi_1,\xi_2) = 0,\ (i = 1,2), \qquad (2.1.14)$$

that is, *the rest frame state* $u(p;\xi_1,\xi_2)$ *is an eigenstate corresponding to zero quanta,* the internal oscillator variables are in their ground state. (As will be shown below this state corresponds to zero spin.)

For the case where the energy is *negative*, eqs. (2.1.13) assert:

$$a_i^+\ u(p,\xi_1,\xi_2) = 0, \quad i = 1,2, \qquad (2.1.15)$$

in other words this eigenstate is to be annihilated by both *creation operators* a_i^+. Since there are no such states, we conclude that *the new Dirac equation describes particles of mass m having only positive energies.*

§2. RELATIONSHIP WITH THE DE SITTER GROUP, SO(3,2)

It is an interesting fact that both the new Dirac equation and the usual Dirac electron equation--as well as the Majorana equation--are all inter-related through the deSitter* group SO(3,2) and in particular to a certain representation of this group. In recognition of this curious relationship, Dirac [DIR 3] has termed this representation, the "remarkable representation".

In order to go more deeply into the structure and properties of the new Dirac equation it is useful to discuss, systematically, the algebraic structures underlying these interrelationships.

We begin by noting that one of the more ingenious algebraic features of the new Dirac equation is the way in which the (numerical) Dirac matrices (of the usual electron equation) play a role via the column operators, Ω, in the new equation. Closer examination of this structure shows that two distinct structural features are involved:

---

* See Note 1 for a brief discussion of the de Sitter group.

(a) A variant of the *Jordan-Schwinger mapping* [JOR 1, SCW 1, BIE 1] (whereby numerical matrices map into boson operators) and

(b) The *tensor operator mapping* [BIE 1], whereby (bilinear) operators map into (numerical) matrices. This latter is, in effect, the inverse to Jordan-Schwinger (boson operator) mapping.

Let us first recall the boson operator mapping. Consider two independent boson operators: $\{a_1, a_1^+ ; a_2, a_2^+\}$ with the commutation relations $[a_i, a_j^+] = \delta_{ij}$, all other commutators vanishing. Consider now a 2 x 2 matrix $A = (A_{ij})$ over the complex numbers. Define the map:

$$J: \quad A = (A_{ij}) \to \sum_{i,j=1}^{2} (a_i^+ A_{ij} a_j), \equiv a^+ A a \qquad (2.2.1)$$

which we will denote by:

$$A \to J(A). \qquad (2.2.2)$$

Then, using the boson commutation relations, it is easily seen that (for 2 x 2 matrices A, B):

$$J([A, B]) = [J(A), J(B)]. \qquad (2.2.3)$$

Expressed in words one sees that the map of the commutator is the commutator of the mappings. That is: *The Jordan-Schwinger mapping preserves commutation relations.*

The mapping used in the new Dirac equation is a more general version of the boson operator mapping, in that both creation and destruction operators occur equivalently. (Thus we have a 4 x 4 matrix mapping for only two bosons.) In order to clarify this more general aspect we first prove an ancillary result.

Lemma: Consider a numerical 4 x 4 matrix $C = (C_{ij})$, $i,j = 1\ldots4$, and let us assume that C is anti-symmetric: $C_{ij} = -C_{ji}$. *Then the mapping:*

$$J : C = (C_{ij}) \to J(C) = \sum_{i,j=1}^{4} Q_i \, C_{ij} \, Q_j \qquad (2.2.4)$$

yields for J(C) *a c-number and not an operator.* (The proof is immediate upon using eq. (2.1.2) and the anti-symmetry. One finds that: $J(C) = \frac{i}{2}\mathrm{tr}\ (\beta C)$ = c-number. In fact, for the 6 linearly independent anti-symmetric 4 x 4 matrices we may choose a basis such that only $J(\beta)$ does not map to zero.

The importance of this elementary result is that only the 10 (linearly independent) *symmetric* 4 x 4 matrices lead to non-trivial operators under the mapping J.

The mapping J, defined in eq. (2.2.4), is, however, not in the most convenient form (since the matrix β will occur awkwardly in commutators). Let us choose to define the *Dirac operator mapping*, denoted by $\mathcal{D}$, in this way:

$$\mathcal{D}: \quad A = (A_{ij}) \to \frac{i}{2} \sum_{k,i,j=1}^{4} Q_k \, \beta_{ki} \, A_{ij} \, Q_j \equiv \frac{i}{2} \tilde{Q}\beta A Q \tag{2.2.5}$$

$$\text{or} \quad A \to \mathcal{D}(A). \tag{2.2.6}$$

If we now *restrict the matrices* A *to be 4 x 4 numerical matrices of the form:* βM (where M is a *symmetric* 4 x 4 matrix) (equivalently, βA is to be symmetric) then we find (after a short calculation) using (2.1.2) that:

$$[\mathcal{D}(A), \mathcal{D}(B)] = \mathcal{D}([A, B]). \tag{2.2.7}$$

For matrices of the restricted form, *the Dirac operator mapping preserves commutation relations.* The Dirac operator mapping thus has the same basic property as the Jordan-Schwinger boson operator mapping, but constitutes a generalization of the J-S map in that the matrices involved are larger (4 x 4 instead of 2 x 2). The price one pays for this generalization (and a price must be paid since the number of *operator* degrees of freedom has not changed!) is that the admissible matrices must have a restricted form.

With these preliminaries accomplished we can now make quite transparent the essence of the structure underlying the new Dirac equation. Let us introduce, as usual, the (Lorentz) covariant γ-matrices to replace the four α-matrices of eqs. (2.1.5) and (2.1.6):

$$\begin{aligned} \gamma_0 &= \beta = i\rho_2 = \begin{pmatrix} 0 & 1 \\ -1 & 0 \end{pmatrix}, \quad & \gamma_1 &= -\rho_3\sigma_3 = \begin{pmatrix} -\sigma_3 & 0 \\ 0 & \sigma_3 \end{pmatrix} \\ \gamma_2 &= \rho_3\sigma_1 = \begin{pmatrix} \sigma_1 & 0 \\ 0 & -\sigma_1 \end{pmatrix}, \quad & \gamma_3 &= -\rho_1 = \begin{pmatrix} 0 & -1 \\ -1 & 0 \end{pmatrix} \end{aligned} \tag{2.2.8}$$

The four matrices $\{\gamma_\mu\}$ form a special representation* of the usual Dirac matrices, but they obey the usual anti-commutation relations:

$$\{\gamma_\mu, \gamma_\nu\} = 2g_{\mu\nu} \tag{2.2.9}$$

---

* This representation is a variant form of the Majorana representation.

with a spacelike metric: $g_{oo} = -1$.

All four $\{\gamma_\mu\}$ are real, as is the fifth one defined by:

$$\gamma_5 = \gamma_0\gamma_1\gamma_2\gamma_3 = i\rho_3\sigma_2 = \begin{pmatrix} i\sigma_2 & 0 \\ 0 & -i\sigma_2 \end{pmatrix}, \quad \gamma_5{}^2 = -1. \tag{2.2.10}$$

Under transposition we have:

$$\gamma_\mu{}^\sim = -\beta\,\gamma_\mu\beta^{-1} \quad , \quad \gamma_5{}^\sim = \beta\gamma_5\beta^{-1} \tag{2.2.11}$$

As a consequence of these symmetry and reality properties we find that the familiar decomposition of the Dirac matrix ring into {Scalar, Vector, Tensor, Axial vector, Pseudoscalor} splits into two sets:

(i) the 10 *symmetric* matrices: (based on V,T)

$$\{\beta\gamma_\mu\} \text{ and } \{\beta[\gamma_\mu,\gamma_\nu]\} \tag{2.2.12a}$$

(ii) the 6 *anti-symmetric* matrices: (based on S,A,P)

$$\{\beta\},\ \{\beta\gamma_5\gamma_\mu\} \text{ and } \{\beta\gamma_5\}\ . \tag{2.2.12b}$$

We see from the form of the 10 symmetric matrices, set (i) above, that they are adapted to the Dirac mapping in such a way that the matrices β drop out. That is: *The Dirac mapping applied to the matrices* $\{\gamma_\mu\}$ *and* $\{[\gamma_\mu,\gamma_\nu]\}$ *preserves commutation relations.*

Let us note that if we define the adjoint:

$$\bar{Q} \equiv \tilde{Q}\,\beta\ , \tag{2.2.13}$$

then the Dirac mapping takes on the suggestive form:

$$\mathcal{D}:\quad A \to \mathcal{D}(A) = \bar{Q}\,A\,Q. \tag{2.2.14}$$

It is now an easy matter to determine that the SO(3,2) group occurs in the symmetry structure of the new Dirac equation, since it is well-known that the 10 Dirac matrices: $\{\frac{i}{2}\gamma_\mu\}$ and $\{\frac{-i}{4}[\gamma_\mu,\gamma_\nu]\}$ obey the commutation relations for the generators of this group. Using the Dirac mapping on these matrices, one finds:

$$\mathcal{D}:\ \tfrac{i}{2}\gamma_\mu \to \mathcal{D}(\tfrac{i}{2}\gamma_\mu) \equiv V_\mu\ , \tag{2.2.15}$$

$$\mathcal{D}:\ \tfrac{-i}{4}[\gamma_\mu,\gamma_\nu] \to \mathcal{D}(\tfrac{-i}{4}[\gamma_\mu,\gamma_\nu]) \equiv S_{\mu\nu}\ . \tag{2.2.16}$$

It is useful to give the operators explicitly in terms of the $\{\xi_i\}$ and $\{\pi_i\}$:

$$S_{12} = \tfrac{1}{2}(\xi_2\pi_1 - \xi_1\pi_2), \quad S_{23} = \tfrac{1}{2}(\xi_1\xi_2 + \pi_1\pi_2), \quad S_{31} = \tfrac{1}{4}(\xi_1^2 + \pi_1^2 - \xi_2^2 - \pi_2^2),$$

$$S_{01} = \tfrac{1}{4}(\xi_1^2 - \pi_1^2 - \xi_2^2 + \pi_2^2), \quad S_{02} = \tfrac{1}{2}(\pi_1\pi_2 - \xi_1\xi_2), \quad S_{03} = \tfrac{1}{2}(\xi_1\pi_1 + \pi_2\xi_2),$$

$$V_1 = \tfrac{1}{2}(\xi_2\pi_2 - \xi_1\pi_1), \quad V_2 = \tfrac{1}{2}(\xi_1\pi_2 + \xi_2\pi_1), \quad V_3 = \tfrac{1}{4}(\xi_1^2 - \pi^2 + \xi_2^2 - \pi_2^2),$$

$$V_0 = \tfrac{1}{4}(\xi_1^2 + \pi_1^2 + \xi_2^2 + \pi_2^2) \,. \tag{2.2.17}$$

All ten of the operators $\{V_\mu, S_{\mu\nu}\}$ in (2.2.17) are Hermitian. It is useful to note that the operator V_0 is positive definite.

It is now easily checked (either directly, or from the commutation relations of the γ-matrices using the map $\mathcal{D}$) that the operators $\{V_\mu, S_{\mu\nu}\}$ obey the commutation relations;

$$[S_{\mu\nu}, S_{\rho\sigma}] = i\,(g_{\mu\rho}S_{\nu\sigma} - g_{\nu\rho}S_{\mu\sigma} + g_{\mu\sigma}S_{\rho\nu} - g_{\nu\sigma}S_{\rho\mu}) \,,$$

$$[S_{\mu\nu}, V_\rho] = i\,(g_{\mu\rho}V_\nu - g_{\nu\rho}V_\mu) \,, \tag{2.2.18}$$

$$[V_\mu, V_\nu] = -i\, S_{\mu\nu} \,.$$

Thus in the space $\mathcal{H}_0$ of an irreducible representation of the relations (2.1.1) the $S_{\mu\nu}$ generate a unitary representation of the homogeneous Lorentz group SO(3,1) (more precisely, of the group SL(2,C)); the V_μ transforms as a four-vector under this representation; and all together they generate a unitary representation of SO(3,2).

If we use indices A, B,... going over 0,1,2,3,5 and a metric tensor g_{AB} with $g_{55} = -1$ (the $g_{\mu\nu}$ are as before and $g_{\mu 5}$ vanish), the entire set of commutation relations (2.2.18) can be compactly written, with $V_\mu = S_{\mu 5}$, in the form:

$$[S_{AB}, S_{CD}] = i\,(g_{AC}S_{BD} - g_{BC}S_{AD} + g_{AD}S_{BC} - g_{BD}S_{CA}) \,. \tag{2.2.19}$$

(Of course, $S_{AB} = -S_{BA}$.)

The SO(3,2) Representation and Its Hilbert Space. The operators $\{V_\mu, S_{\mu\nu}\}$ obtained from the Dirac mapping are quadratic in the oscillator variables ξ_i and π_i. To obtain a concrete description of the Hilbert space $\mathcal{H}_o$ which carries the representation generated by the operators $\{V_\mu, S_{\mu\nu}\}$ we may realize the ξ_i as diagonal and the π_1 by: $\pi_1 = -i\, \partial/\partial\xi_i$. The vectors in $\mathcal{H}_o$ are then square integrable functions of $\{\xi_i\}$. Another concrete description of H_0 arises on introducing the boson operators:

$$a_j = \frac{1}{\sqrt{2}}\,(\xi_j + i\pi_j)\ ,\quad a_j^+ = \frac{1}{\sqrt{2}}(\xi_j - i\pi_j),\quad j = 1,2 \tag{2.2.20}$$

which obey the standard commutation relations

$$[a_j, a_k] = [a_j^+, a_k^+] = 0\ ,\quad [a_j, a_k^+] = \delta_{jk}\ , \tag{2.2.21}$$

We can, in the standard way, realize an orthonormal basis of ket vectors in H_o. These ket vectors are simultaneous eigen-kets of the (two) number operators: $N_i = a_i^+ a_i$, and are explicitly:

$$|n_1, n_2> = (n_1!\, n_2!)^{-\frac{1}{2}}(a_1^+)^{n_1}(a_2^+)^{n_2}\,|\,0, 0>\ ,$$

$$a_j|0,0> = 0\ ,\quad n_1, n_2 = 0,1,2, \ldots\ . \tag{2.2.22}$$

This is at most a two valued representation.

Let us now make explicit the representation generated by the operators $\{V_\mu, S_{\mu\nu}\}$. A general element g in SO(3,2) is represented by a unitary operator U(g) acting on H_o. Denoting a general element in g by the 10 real parameters $\omega^{AB} = -\omega^{BA}$ (canonical coordinates of the first kind) we may write:

$$g \in SO(3,2):\quad U(g) = \exp \frac{i}{2}\,\omega^{AB}\, S_{AB}\ . \tag{2.2.23}$$

This is a representation since:

$$U(g_1)\ U(g_2) = U(g_1 g_2) = U(g_{12}). \tag{2.2.24}$$

The representation: $g \to U(g)$ is unitary since the ω^{AB} are real and the S_{AB} are Hermitian: $(U(g))^+ = (U(g))^{-1} = U(g^{-1})$.

On restricting g to the elements A of the homogeneous Lorentz subgroup SO(3,1) in SO(3,2), we get a unitary representation of SO(3,1) by operators U(A) generated by $S_{\mu\nu}$. Both representations, U(g) of SO(3,2) and U(A) of SO(3,1), are reducible and in each case we have a direct sum of two unitary irreducible representations. This can be seen as follows. The Hilbert space H_0 splits into two mutually orthogonal subspaces H_+, H_- say, spanned by states $|n_1,n_2>$ with $n_1 + n_2$ even and odd respectively. It is clear that $H_+(H_-)$ contains only integer (half odd integer) spin representations of the angular momentum algebra (2.19), each such spin value occurring just once. Moreover all ten operators S_{AB} leave the subspaces $H\pm$ invariant, thus accomplishing the reduction of the representations U(g) [U(Λ)] of SO(3,2) [SO(3,])] to irreducible ones. The unitary irreducible representations of SO(3,1) appearing here are the two well-known Majorana ones [MAJ 1], those of SO(3,2) are what have been termed "remarkable representations" by Dirac [DIR 3]. These representations are distinguished by the existence of a large number of algebraic relationships [BOH 1] among their generators: we have, for example,

$$S^{\mu\nu}S_{\mu\nu} = \frac{-3}{2}, \quad V^{\mu}V_{\mu} = \frac{1}{2} \quad ,$$
$$\varepsilon^{\mu\nu\rho\sigma} S_{\mu\nu}S_{\rho\sigma} = \varepsilon^{\mu\nu\rho\sigma} S_{\mu\nu}V_{\rho} = 0 \ . \tag{2.2.25}$$

An 'Inverse' to the Dirac Mapping. We have seen that an essential aspect in the use of the column operator Q is the existence of the Dirac mapping, carrying the 4 x 4 matrices $\{\gamma_\mu, [\gamma_\mu, \gamma_\nu]\}$ into the operators S_{AB}. The question naturally arises: can one go backwards, and recover the 4 x 4 matrices from the S_{AB}? The answer is yes, and this 'inverse' mapping is the *tensor operator relation.*

To see this consider the 4 x 1 (column vector) operator Q, and the commutator: $[S_{AB}, Q]$.

This commutator is easily evaluated and one finds:

$$[S_{AB}, Q] = - i \Sigma_{AB} Q, \tag{2.2.26}$$

where:
$$\Sigma_{\mu\nu} = \frac{-i}{4}[\gamma_\mu, \gamma_\nu] \quad , \tag{2.2.27a}$$
$$\Sigma_{\mu 5} = \frac{i}{2} \gamma_\mu \ . \tag{2.2.27b}$$

Thus we see that the tensor operator relation, eq. (2.2.26), effects mapping:

$$Q: \quad S_{AB} \rightarrow \Sigma_{AB} \ , \tag{2.2.28}$$

which is *inverse* to the Dirac operator mapping:

$$\mathcal{D}: \quad \Sigma_{AB} \rightarrow S_{AB} \quad .$$

Remark: It may not be clear why eq. (2.2.26) is called the "tensor operator relation". To see this, recall that for the angular momentum operator $\vec{J} = \{J_q\}$, and the tensor operator $T_{j,m}$ the standard tensor operator relation reads:

$$\left[J_q, T_{j,m}\right] = \sum_{m'} C^{j\ 1\ j}_{m\ q\ m'} \ T_{j,m'} \quad .$$

The relation is precisely in the form of eq. (2.2.26) with the operator S_{AB} playing the role of the angular momentum operator $\vec{J}$ and the matrix Σ_{AB} the role of the Wigner coefficients $C^{j1j}_{mqm'}$.

It is now obvious from the construction that Σ_{AB} satisfy exactly the same commutation relations as the S_{AB}, that is, eq. (2.2.19).

The tensor operator relation, eq. (2.2.26) has an integrated form which reads:

$$g \in SO(3,2): \quad U(g)\; Q\; U(g)^{-1} = S(g^{-1})Q. \qquad (2.2.29)$$

Here $S(g)$ is a real 4 x 4 non-unitary matrix representation (two-valued) of $SO(3,2)$. Namely we have $S(g) = \exp(\frac{1}{2}\,\omega^{AB}\,\Sigma_{AB})$ and $S(g')S(g) = S(g'g)$. In particular, if we restrict L to elements Λ of the Lorentz subgroup, $SO(3,1)$, we encounter the matrix $S(\Lambda)$ which is precisely the matrix by which the 4-component spinor in the usual Dirac (electron) equation transforms.

Let us emphasize that the Dirac mapping, and its inverse, although preserving the commutation relations, *did not preserve the Hermitian character of the generators*, as the fact that $U(g)$ is unitary, while $S(g)$ is not, verifies.

There are analogous relations, similar to eq. (2.2.29), for the generators $\{V_\mu, S_{\mu\nu}\}$. For Lorentz transformations these read:

$$U(\Lambda)\; V_\mu U(\Lambda)^{-1} = \Lambda^\rho{}_\mu V_\rho \;, \; U(\Lambda)\; S_{\mu\nu}\; U(\Lambda)^{-1} = \Lambda^\rho{}_\mu \Lambda^\sigma{}_\nu S_{\rho\sigma} \;. \qquad (2.2.30)$$

It is useful to record also the analogous relations for the matrices β and γ_μ. We find:

$$S(g)^{\sim}\; \beta S(g) = \beta \;, \quad S(\Lambda)^{-1}\; \gamma_\mu S(\Lambda) = \Lambda_\mu{}^\rho \gamma_\rho \;. \qquad (2.2.31)$$

Why the Matrices are Real in the New Dirac Equation. In the discussion of the new Dirac equation in §1, the reality of the representation of the matrices was stated as necessary, but unproved. It is now easy to see why this condition must hold.

If we examine the representation $S(g)$ of the group $SO(3,2)$ we see that it is *real* as well as non-unitary. Thus when Q is acted upon by $U(g)$, as in eq. (2.2.29), it undergoes the real, non-unitary, linear transformation $S(g)$, exactly as a real four-component spinor would.

Turning this result around, we see that *because* Q is a *real* four-component spinor in the new Dirac equation it is necessary that the transformation be real, and this requires the γ-matrices in turn to be real. This transformation law of Q under $SO(3,1)$ (more properly $SL(2,C)$) is compactly conveyed through the statement that the two-component operator

$$\begin{pmatrix} \xi_1 + i\xi_2 \\ \pi_1 - i\pi_2 \end{pmatrix} \qquad (2.2.32)$$

behaves as an (undotted) two component SL(2,C) spinor.

§3. SOME PROPERTIES OF THE NEW DIRAC EQUATION

Poincaré Covariance: We have now assembled in §2 all the technical apparatus needed to deal efficiently with the transformation properties of the new Dirac equation. Let (d,Λ) be an element of the Poincaré group corresponding to the space-time transformations:

$$(d,\Lambda): \quad x^\mu \to (x')^\mu \equiv \Lambda^\mu{}_\nu x^\nu + d^\mu \quad . \qquad (2.3.1)$$

The new Dirac equation reads:

$$(\gamma^\mu \partial_\mu + m)\ Q\psi(x) = 0 \ . \qquad (2.3.2)$$

In this equation $\psi(x)$ is regarded as a function of x and as a vector in the Hilbert space spanned by the "internal" degenerate oscillator variables. Instead of the four component spinor functions of the electron equation we have here a function with an infinite number of components. We do not write these components, however, as they would get mixed up with the four components of ψ. When one applies a Poincaré transformation $(d,\Lambda)^*$, $\psi(x)$ transforms according to

$$\psi'(x) = U(\Lambda)\ \psi(\Lambda^{-1}(x-d)), \qquad (2.3.3)$$

where $U(\Lambda)$ is defined on the Hilbert space of internal oscillator variables and given by (2.2.23) (restricted to SO(3,1)). The Poincaré covariance of equation (2.3.2) will be established by showing that the transformed wave function (2.3.3) satisfies (2.3.2) for every (d,Λ) i.e.

$$(\gamma^\mu \partial_\mu + m)\ Q\psi'(x) = 0 \qquad (2.3.4)$$

for all (d,Λ) with $\psi'(x)$ given by (2.3.3). The covariance for space time translations (d,1) is obvious so that we may restrict ourselves to showing (2.3.4) for Lorentz transformation $(0,\Lambda)$. In order to do this let us first introduce two auxiliary results. First, note (2.2.29)

---

* Actually we need to consider the covering group (d,Λ) where Λ is a 2 x 2 complex matrix with unit determinant SL(2,C). We shall do this later.

which implies

$$U^{-1}(\Lambda)\ QU(\Lambda)\ =\ S(\Lambda)\ Q. \tag{2.3.5}$$

Second, recall the analogous result for the γ matrices:

$$S^{-1}(\Lambda)\gamma^{\mu}\ S(\Lambda)\ =\ \Lambda^{\mu}{}_{\nu}\gamma^{\nu}\ . \tag{2.3.6}$$

Using these results we have with (2.3.5)

$$(\gamma^{\mu}\partial_{\mu} + m)\ Q\psi'(x)\ \equiv\ (\gamma^{\mu}\partial_{\mu} + m)QU(\Lambda)\psi(\Lambda^{-1}x)$$

$$= (\gamma^{\mu}\partial_{\mu} + m)U(\Lambda)\ U^{-1}(\Lambda)QU(\Lambda)\ \psi(\Lambda^{-1}x) =$$

$$= (\gamma^{\mu}\partial_{\mu} + m)\ U(\Lambda)\ S(\Lambda)\ Q\psi(\Lambda^{-1}x) =$$

$$= U(\Lambda)\ (\gamma^{\mu}\partial_{\mu} + m)\ S(\Lambda)\ Q\psi(\Lambda^{-1}x).$$

Next, introducing

$$y = \Lambda^{-1}x \tag{2.3.7}$$

and using (2.3.6) which gives

$$S^{-1}(\Lambda)\gamma^{\mu}\partial_{\mu}S(\Lambda)\ \equiv\ S^{-1}(\Lambda)\gamma^{\mu}\ \frac{\partial}{\partial x^{\mu}}\ S(\Lambda)\ =$$

$$= \gamma^{\mu}\ \frac{\partial}{\partial y^{\mu}}$$

one has

$$(\gamma^{\mu}\partial_{\mu} + m)Q\psi'(x)\ =\ U(\Lambda)\ S(\Lambda)\ (\gamma^{\mu}\frac{\partial}{\partial y^{\mu}} + m)\ Q\psi(y)\ =\ 0,$$

where one has used the validity of (2.3.2) for all x.

For future use note

$$U(\Lambda)\ (\gamma^{\mu}\partial_{\mu} + m)\ Q\ U^{-1}(\Lambda)\ =$$

$$= S^{-1}(\Lambda)(\Lambda^{-1\mu}{}_{\rho}\ \gamma^{\rho}\partial_{\mu} + m)\ Q. \tag{2.3.8}$$

The analogy to the electron equation is obvious. Analogous to the Dirac scalar field $\left(\bar{\psi}(x)\psi(x)\right)_{\text{Dirac}} \equiv \left(\psi^{+}(x)\gamma^{0}\psi(x)\right)_{\text{Dirac}}$ one here has the scalar field

$$\psi^+(x)\ \psi(x). \tag{2.3.9}$$

That (2.3.9) is a scalar field follows from (2.3.3). Integrating (2.3.9) over all space time gives an invariant, but as particles persist in time this is not a satisfactory norm for the states $\psi(x)$. Just as in the electron equation case the inner product and norm of the states is based on a conserved current.

Conserved Current. By multiplying the new Dirac equation (2.3.2) in front by $Q^+\gamma^o$ and then using $Q^+\gamma^o\gamma^\mu Q = V^\mu$, $Q^+\gamma^o Q = 1$ one finds that $\psi(x)$ satisfies the Majorana equation $(V^\mu\partial_\mu + m)\psi(x) = 0$. From this it follows that the current

$$j^\mu(x) = \psi^+(x)\ V^\mu\psi(x) \tag{2.3.10}$$

is conserved:

$$\partial_\mu j^\mu(x) = 0. \tag{2.3.11}$$

That $j^\mu(x)$ is a vector field follows from (2.3.3) and (2.2.30). All this allows one to write the invariant norm of a state $\psi(x)$ as

$$||\psi||^2 = \int d^3x\ \psi^+(x)\ V^0\psi(x). \tag{2.3.12}$$

Physical inner product. By a similar argument the inner product of two states $\psi(x)$, $\phi(x)$, each a solution of (2.3.2) and transforming with (2.3.3), is given invariantly by

$$(\psi,\phi) = \int d^3x\psi^+(x)\ V^0\ \phi(x). \tag{2.3.13}$$

So far we have considered $\psi(x)$ to be a function of x and a vector in the internal Hilbert space of two degenerate oscillators. On a basis with ξ_1 and ξ_2 diagonal the vector $\psi(x)$ has as components the wave function $\psi(x;\xi_1,\xi_2)$, in terms of this wave function (2.3.10) reads

$$j^\mu(x) = \int d\xi_1 d\xi_2\ \psi^*(x;\xi_1,\xi_2)V^\mu\psi(x_1;\xi_1,\xi_2) \tag{2.3.10a}$$

and the inner product of two wave functions ψ,ϕ is given by

$$(\psi,\phi) = \int d^3x\ d\xi_1 d\xi_2\ \psi^*(x_1;\xi_1,\xi_2)V^0\ \phi(x_1;\xi_1,\xi_2). \tag{2.3.13a}$$

To be entirely clear let us repeat the invariance argument with a single ket vector $|\psi\rangle$ defined in the direct product Hilbert space instead using the mixed Hilbert space--wave function $\psi(x)$. The equation (2.3.2) in this new language reads

$$(i\gamma^{\mu}P_{\mu} + m)\ Q|\psi\rangle = 0. \tag{2.3.2a}$$

The transformation (2.3.3) then reads

$$|\psi'\rangle = \mathcal{U}\ |\psi\rangle \tag{2.3.3a}$$

with $\mathcal{U}$ in the direct product Hilbert space and is a product of $U(\Lambda)$ of (2.3.3) with a unitary transformation which changes P to P'.

To establish the covariance of (2.3.2a) one must show

$$(i\gamma^{\mu}P_{\mu} + m)\ Q\mathcal{U}|\psi\rangle = 0. \tag{2.3.4a}$$

This is done by calculating

$$\begin{aligned} & U^{-1}(i\gamma^{\mu}P_{\mu} + m)\ QU = \\ & = (i\gamma^{\mu}U^{-1}P_{\mu}U + m)\ U^{-1}QU = \\ & = (i\gamma^{\mu}P'_{\mu} + m)\ S(\Lambda)Q = \\ & = S(\Lambda)\ S^{-1}(\Lambda)(i\gamma^{\mu}P'_{\mu} + m)S(\Lambda)Q = \\ & = S(\Lambda)(i\gamma'^{\mu}P'_{\mu} + m)Q = \\ & = S(\Lambda)(i\gamma^{\mu}P_{\mu} + m)Q. \end{aligned} \tag{2.3.14}$$

Equation (2.3.14) establishes (2.3.4a) as well as the fact that the operator $(i\gamma^{\mu}P_{\mu} + m)Q$ *transforms as a spinor.*

General Solution of Dirac's New Equation. In §1 we obtained the rest frame solution to Dirac's new equation, and it is useful now to develop--completely explicitly--the general solution for an arbitrary four-momentum. Let us assume a planewave solution of the form:

$$\Psi_p(x,\xi) = e^{ip\cdot x}u(p,\xi). \tag{2.3.15}$$

Here $p = (p_{\mu})$ is a numerical positive time-like four-vector obeying $p\cdot p = -m^2$, and $u(p,\xi)$ is a wave-function realized in H_o ie, (p,ξ) are taken diagonal. It follows, using (2.3.2), that $u(p,\xi)$ must obey the four equations:

$$(i\gamma\cdot p + \frac{mc}{\hbar})\, Q\, u(p,\xi) = 0. \tag{2.3.16}$$

These four equations become the two partial differential equations: (using the boson operator term of Q, cf. p. 10)

$$-i\frac{\partial}{\partial q_1}u(p;\xi_1\xi_2) = (p^o - p^3)^{-1}(p_1 + im)q_1 - p_2q_2)u(p,\xi_1\xi_2),$$

$$-i\frac{\partial}{\partial q_2}u(p;\xi_1\xi_2) = (p^o - p^3)^{-1}((-p_1 + im)q_2 - p_2q_1)u(p;\xi_1\xi_2), \tag{2.3.17}$$

and two other equations equivalent to these since $p^2 = -m^2$. Up to a factor the solution is

$$u(p,\xi_1\xi_2) = \exp\left[\frac{i}{(p^o - p^3)}\{\tfrac{1}{2}p_1(\xi_1{}^2 - \xi_2{}^2) - p_2\xi_1\xi_2 + \tfrac{i}{2}m(\xi_1{}^2 + \xi_2{}^2)\}\right].$$

The complete plane wave solution $\Psi_p(x;\xi_1\xi_2)$, normalized to a delta function in the 3-momentum p according to the physical inner product (2.3.13), is found to be:

$$\Psi_p(x;\xi_1\xi_2) = \frac{m}{2\pi^2\hbar}\,[p^o(p^o - p^3)]^{-1/2}\cdot e^{ip\,x}\,. \tag{2.3.18}$$

$$\exp\left[\frac{i}{\hbar(p^o - p^3}\ \{\tfrac{1}{2}p_1(\xi_1{}^2 - \xi_2{}^2) - p_2\xi_1\xi_2 + \tfrac{i}{2}m(\xi_1{}^2 + \xi_2{}^2)\}\right];$$

$$(\Psi_{p'},\Psi_p)_{\text{phys.}} \equiv \int d^3x \int_{-\infty}^{\infty} dq_1 \int_{-\infty}^{\infty} dq_2\ \Psi_{p'}(x;\xi_1\xi_2)^* V_o \Psi_p(x;\xi_1\xi_2)$$

$$= \delta^3(\vec{p}' - \vec{p}).$$

§4. AN ALTERNATIVE, MORE ILLUMINATING, VIEW OF THE STRUCTURE

We have obtained the general solution to the new Dirac equation, in §3, by choosing a particular realization of the operators and directly solving the resulting differential equations. A more illuminating view of the structure of this general solution results by recognizing that this solution can be obtained simply by Lorentz transforming the rest frame solution, eqs. (2.1.12,14), and this in turn leads to the recognition that the new Dirac equation is no more and no less than *a covariant way to assert that the internal oscillator structure is in a state of zero quanta*. In other words, the new Dirac system is an ingenious realization of a positive energy, time-like, relativistic harmonic oscillator and hence an object of considerable theoretical usefulness. The present section is devoted to making this basic structure understandable, and in Chapter 3, we will further extend the structure to include super-symmetry.

The Concept of Aligned Bosons. We begin by noting that the operator Ω, according to eq. (2.2.29), transforms under Lorentz transformations as a spinor. It follows that the boson operators, eq. (2.2.20),--since they are constructed linearly from Ω-- also undergo an induced Lorentz transformation.

Under a general Lorentz transformation, Λ, we find for the boson operators that:

$$\Lambda: \quad a_j \to a'_j = U(\Lambda)\, a_j\, U(\Lambda)^{-1}, \qquad a_j^+ \to (a'_j)^+ = U(\Lambda)\, a_j^+\, U(\Lambda)^{-1} . \tag{2.4.1}$$

It is clear that the boson commutation relations are preserved under this transformation.

It is also clear that the vacuum ket, on which the transformed bosons act, *must have changed* under the Lorentz transformation Λ.

The vacuum ket, $|0\rangle$, is defined by two conditions: $a_i|0\rangle = 0$, $i = 1,2$. It is invariant for those Lorentz transformations which leave the time axis invariant; that is, the rotations leaving the unit four-vector (0 0 0 1) invariant. For the sake of precision, one includes this information in the notation for the vacuum ket. Let us denote the pure Lorentz transformation Λ that takes (0 0 0 m) into a general four-momentum p by B_p . We can label B_p uniquely by the time-like *unit* four-vector $\hat{p}$, since there are only three parameters involved and all momenta are time-like.

Accordingly we denote the vacuum ket by: $|0;\hat{p}\rangle$ with the rest frame vacuum ket being $|0;\hat{p}_o\rangle$, $\hat{p}_o = (0\ 0\ 0\ 1)$. Thus we have:

$$B_p \;:\; |0;\hat{p}_o\rangle \to |0;\hat{p}\rangle = U(\Lambda(p))\,|0;\hat{p}_o\rangle \quad . \tag{2.4.2}$$

Consider now the operator, Π, defined by:

$$\Pi \equiv \tfrac{1}{2}(\mathbb{1} + i\gamma \cdot \hat{P}), \qquad (\hat{P})_\mu \equiv m^{-1}P_\mu \quad . \tag{2.4.3}$$

This is a projection operator, since $\Pi^2 = \Pi$ follows from $P\cdot P = -m^2$. (Note that P is the four-momentum operator, and observe that both eq. (2.4.2) and eq. (2.4.1) are meaningful if the transformation is operator valued in P.)

In terms of this more explicit notation, we now denote the transformed bosons of eq. (2.4.1) by:

$$B_p \;:\; a_j(\hat{P}_\circ) \to a_j(\hat{P}) \;, \qquad a_j^+(\hat{P}_\circ) \to a_j(\hat{P}). \tag{2.4.1'}$$

We call these transformed bosons "aligned bosons".

The new Dirac equation now takes the form:

$$T_a \;|\Psi\rangle = 0, \; (a = 1, \ldots, 4), \tag{2.4.4}$$

where:

$$T_a \equiv (\Pi\Omega)_a = (\tfrac{1}{2}(\mathbb{1} + i\gamma \cdot \hat{P})\;\Omega)_a. \tag{2.4.5}$$

The operator T_a, because of the projection operator Π in its definition, has only two independent components. In a frame where the projection operator Π is diagonal, these components are precisely $a_1(\hat{P})$ and $a_2(\hat{P})$, as given by eq. (2.4.1'). Thus we see that an equivalent form of eq. (2.4.4) is:

$$a_i(\hat{P}) \;|\Psi\rangle = 0. \tag{2.4.6}$$

In other words, *the new Dirac equation is simply the assertion that the ket* $|\Psi\rangle$ *corresponds to the state of zero quanta for the aligned boson operators.* In particular, for the rest frame, we recover eqs. (2.1.13).

The covariant creation operators corresponding to the covariant (destruction) operators T_a eq. (2.4.4) can be defined using the orthogonal projection operator, $\beta^{-1}\Pi^+\beta$. These take the form:

$$\overline{T} = \tfrac{1}{2}(\mathbb{1} - i\gamma \cdot \hat{P})\;\Omega, \tag{2.4.7}$$

or alternatively:

$$\overline{T} = T^+ \;\gamma^0. \tag{2.4.8}$$

For later use we note the commutation relations:

$$[T_a, T_b] = 0 = [\overline{T}_a, \overline{T}_b] \;, \tag{2.4.9}$$

$$[\overline{T}_a, T_b] = (-i)\tfrac{1}{2}\left(\mathbb{1} + i\hat{p}\cdot\gamma\right)_{ab} \;. \tag{2.4.10}$$

It is not hard to compute the effect of $U(\Lambda(p))$ on a_j explicitly, especially since we are dealing with a pure Lorentz transformation and we know that the two-component object (2.2.32) behaves as a complex spinor under SL(2,C). In this way we get

$$a_1(\hat{p}) = \frac{(p^o + m)}{[2m(p^o + m)]^{-\frac{1}{2}}}\left[a_1 + (p_3 - ip_1)\, a_1^\dagger + ip_2 a_2^\dagger\right],$$

$$a_2(\hat{p}) = \frac{(p^o + m)}{[2m(p^o + m)]^{-\frac{1}{2}}}\left[a_2 + (p_3 + ip_i)\, a_2^\dagger + ip_2 a_1^\dagger\right]. \tag{2.4.11}$$

(The creation operators, $a_i^+(\hat{p})$, are the Hermitian adjoints to the $a_i(\hat{p})$ above.) A part from a constant, u(p) is the unique vector in H_o annihilated by $a_i(\hat{p})$ and $a_2(\hat{p})$.

§5. GENERALIZATION TO NON-ZERO SPIN STATES

Once we have recognized the relation between the new Dirac equation and the quanta of the aligned boson operators, as discussed in §4 above, we are in a position to generalize immediately to higher spin states.

To do so it is useful first to recall now the Jordan-Schwinger mapping allows one to realize any angular momentum multiplet in terms of the two boson creation operators a_1 and a_2. Defining the angular momentum operator, J_i, as the Jordan-Schwinger map of the Pauli matrices, that is, (see eq. 2.2.1):

$$J_i = J\left(\frac{1}{2}\sigma_i\right), \tag{2.5.1}$$

one finds that a realization of the angular momentum multiplet {j,m}, (m - j, j-1,..., -j with j = 0, 1/2, 1,...) is given by the eigen-kets:

$$|jm> \equiv [(j+m)!(j-m!]^{-\frac{1}{2}}\, (a_1^+)^{j+m}\, (a_2^+)^{j-m}\, |0> , \tag{2.5.2}$$

which obey, by construction (using the J-S map) the relations:

$$J_3|jm> = m|jm>$$

and

$$J^2|jm> \equiv (J_1^2 + J_2^2 + J_3^2)|\, jm> = j(j+1)\, |jm> . \tag{2.5.3}$$

In order to make use of this familiar construction we first note that the *aligned boson operators* $a_i^+(\hat{P})$, of eq. (2.4.1') are to replace the boson operators of the J-S map and secondly that the desired angular momentum operators in terms of these aligned boson operators are to be obtained from the Dirac mapping (using the aligned Q) of the matrices Σ_{ij}, eq. (2.2.27), which generate the rotation sub-group of SO(3,2).

Thus the use of aligned bosons allows one to define the operators $S_k(\hat{P}) = \frac{1}{2}e_{jk}S_{ij}(\hat{P})$; these operators obey the commutation relations for angular momentum.

Consider now an eigen-ket of the momentum operator P having a time-like positive energy momentum p:

$$P_\mu |p\rangle = p_\mu |p\rangle ,$$

$$p \cdot p = -m^2, \quad p_0 > 0. \tag{2.5.4}$$

For such an eigen-ket, the ground state of the internal system is defined by the eigen-ket $|0;\hat{p}\rangle$ (see eq. 2.4.2) corresponding to zero quanta for the aligned bosons:

$$a_i(\hat{p}) |\, 0; \; \hat{p}\rangle = 0, \; i = 1,2. \tag{2.5.5}$$

[Note that the 'operator-valued' aligned boson $a_i(\hat{P})$ when acting on the eigen-ket $|p\rangle$ of the product ket: $|p\rangle|0;\hat{p}\rangle$ goes to the 'numerical-valued' aligned boson $a_i(\hat{p})$. Similarly the angular momentum operators $S_k(\hat{P})$ acting on the momentum eigen-ket become $S_k(\hat{p})$.]

We now recognize that the operators $S_k(\hat{p})$ are the generators of the "little group" (stability group) for eigenstates of the system having momentum p. Taking p to be the momentum in the rest frame $p_o = (000\ m)$ we recover earlier results that the rest frame Dirac solution (having the eigen-ket $|0;\hat{p}\rangle$ with no quanta) is invariant under rotations generated by $S_k(\hat{P}_o) = S_k$.

With this insight into the meaning of the Dirac's result we can now give explicitly the eigen-kets for the system characterized by the time-like, positive energy, momentum p and an arbitrary internal spin j, m. Such an eigen-ket has the form:

$$|p; \; j,m\rangle \equiv [(j+m)!(j-m)!]^{-1/2} \cdot$$

$$(a_1^+(\hat{p}))^{j+m} (a_s^+(\hat{p}))^{j-m} \; |0; \; \hat{p}\rangle \,|\, p\rangle \; . \tag{2.5.6}$$

Let us verify that the eigen-kets given by eq. (2.5.6) do indeed satisfy the conditions stated. To show that the momentum is correct, we operate on eq. (2.5.6) with P and verify that: $P \to p$, (recalling that the eigenvalue was defined by eq. (2.5.4) to be time-like with positive energy). Next consider a general Lorentz transformation Λ restricted by the requirement that p be invariant under Λ, that is: $\Lambda(p) = p$. Let us denote be B_p the pure Lorentz transformation (a boost) that takes the rest frame momentum vector $p_o = (000\ m)$ into p. Then from $\Lambda(p) = p$ we see that $B_p^{-1} \Lambda B_p$ leaves the vector p_o invariant, and thus is a rotation, R:

$$B_p^{-1} \Lambda B_p = R. \tag{2.5.7}$$

Using this display information we can now determine how the eigenket $|p;\ jm\rangle$ transforms under Λ. First we note that, from eq. (2.5.6), we have:

$$|p;\ jm\rangle = \mathcal{U}(B_p)\,|p_o;\ jm\rangle\ . \tag{2.5.8}$$

Under the transformation Λ we find:

$$\Lambda:\ |p;\ jm\rangle \to |p;\ jm\rangle' = \mathcal{U}(\Lambda)|p;\ jm\rangle \tag{2.5.9}$$

$$= \mathcal{U}(\Lambda)U(B_p)\ |p_o;\ jm\rangle,\ \text{(using 2.5.8)}$$

$$= \mathcal{U}(B_pRB_p^{-1})\,\mathcal{U}(B_p)\ |p_o;\ jm\rangle\ ,\ \text{(using 2.5.7)}$$

$$= \mathcal{U}(B_p)\,\mathcal{U}(R)\ |p_o;\ jm\rangle\ .$$

The rotation R has, however, a matrix action on vectors in the rest frame, so that:

$$\mathcal{U}(R)|p_o;\ jm\rangle = \sum_{m'} D^{(j)}_{m'm}(R)\ |p_o;\ jm'\rangle\ , \tag{2.5.10}$$

where $D^j_{m',m}(R)$ is the rotation matrix for angular momentum j.

Thus we find that under Lorentz transformation, Λ, which leaves p invariant, the ket vectors $|p;\ jm\rangle$ transform by a little group rotation with a rotation matrix appropriate to angular momentum j:

$$\Lambda:\ |p;\ jm\rangle \to |p;\ jm\rangle' = \sum_{m'} D^j_{m'm}(R)\ |p;\ jm'\rangle\ . \tag{2.5.11}$$

where R is the little group rotation of eq. (2.5.7).

The properties expressed by eqs. (2.5.11) and (2.5.4) suffice to demonstrate that the ket vectors of eq. (2.5.6) are Poincare eigen-kets belonging to the irrep labelled by mass m and spin j.

It should be noted that this explicit realization of the set of all Poincaré irreps (m,j) using aligned bosons is a very economical one, and realizes the Wigner construction of these states in a compact and uniform way.

The construction is, however, purely kinematical and as yet no wave equation--such as Dirac's new wave equation (which picks out $j = 0$) --has been given. It is possible to give wave equations, which are generalizations of Dirac's new equation, having a given mass m and spin j as eigenstates [see BIE 2.] However, these involve progressively higher order derivatives with repect to x and since in any case we will not need such results below, these equations will be omitted.

Polarization operators: The fact that the stability group operators $S_k(\hat{P})$ can be explicitly given in terms of aligned bosons is of practical importance when reactions involving polarized particles are considered.

The standard way to treat polarization for a relativistic particle of spin j and mass m [see MIC 1] is to introduce the Pauli-Lubansky operator, W, defined by:

$$W_\mu = e_{\mu\nu\lambda\sigma} P^\nu M^{\lambda\sigma} \quad . \tag{2.5.11}$$

The operators P and W commute, but the components of W do not commute, obeying instead:

$$[W_\alpha, W_\rho] = ie_{\alpha\rho\nu\sigma} P^\gamma W^\delta \quad . \tag{2.5.13}$$

The two Poincare invariants are given by:

$$P.P \rightarrow m^2$$

$$W.W \rightarrow m^2 j(j + 1). \tag{2.5.14}$$

In order to define polarization operators in the usual way one introduces a *tetrad*, a set of four oriented orthonormal four-vectors: $\{n^{(\alpha)}(p)\}$, $(\alpha = 0,1,2,3)$ where:

$$n^{(\alpha)} . n^{(\alpha')} = g^{\alpha\alpha'} \quad , \tag{2.5.15}$$

$$e^{\lambda\mu\nu\sigma} n_\lambda^{(\alpha)} n_\mu^{(\beta)} n_\nu^{(\gamma)} n_\sigma^{(\delta)} = e^{\alpha\beta\gamma\delta} \quad . \tag{2.5.16}$$

One then defines four-vector polarization operators $W^{(\alpha)}$ with respect to the tetrad coordinates:

$$W^{(\alpha)} = n^{(\alpha)} . W. \tag{2.5.17}$$

Choosing the tetrad vector $n^{(o)}$ to be along the momentum itself, $n^{(o)} = m^{-1}p$, one finds $W^{(o)} = 0$. The three=vectors $W^{(i)}/m$ (i = 1,2,3) are the desired polarization operators.

By contrast to this circuitous procedure, we see that the three operators: $S_k(\hat{P})$ *are precisely the desired polarization operators, defined abstractly, without the artifice of a tetrad construction.* (Note that the $\{S_k(\hat{P})\}$ obey angular momentum commutation relations, generate little group rotations, and have eigenvalues $\vec{S}(\hat{P}).\vec{S}(P) \to j(j+1)$, $S_3(\hat{P}) \to m$. These are the desired characteristics.)

§6. MINIMAL ELECTROMAGNETIC INTERACTION IS FORBIDDEN FOR THE NEW DIRAC EQUATION

Let us now demonstrate why it is that for the new Dirac equation interaction with an external electromagnetic field via minimal coupling is not possible. Let us consider the vector potential A_μ and the field strengths,

$$F_{\mu\nu}(x) = \partial_\mu A_\nu(x) - \partial_\nu A_\mu(x), \tag{2.6.1}$$

and consider the equation obtained from (2.1.6) by the minimal replacement rule:

$$(\gamma_\mu \Pi^\mu + m)\,\Omega\,\psi(x) = 0, \tag{2.6.2}$$

where

$$\Pi^\mu = \partial^\mu - \frac{ie}{\hbar} A^\mu .$$

Premultiplying this equation for ψ by the matrix operator $(\gamma\cdot\Pi - m)$ we get, by the usual properties of the γ-matrices,

$$\left(\Pi^2 - m^2 - \frac{ie}{4} F^{\mu\nu}\,[\gamma_\mu,\gamma_\nu]\right)\Omega\,\psi = 0. \tag{2.6.3}$$

This result implies new algebraic conditions on ψ not involving space-time derivatives at all. To see this, apply $\Omega^+\beta\gamma_5$ to eq. (2.6.3) on the left, to obtain:

$$F^{\mu\nu}\,\Omega^+\beta\gamma_5[\gamma_\mu,\gamma_\nu]\Omega\,\psi = 0 . \tag{2.6.4}$$

However, since

$$\gamma_5[\gamma_\mu,\gamma_\nu] = \varepsilon_{\mu\nu\rho\sigma}\,\gamma^\rho\gamma^\sigma, \quad \varepsilon_{0123} = +1. \tag{2.6.5}$$

we can express eq. (2.6.4) in the form:

$$F^{*\mu\nu} S_{\mu\nu}\psi = 0, \tag{2.6.6}$$

(Here the tensor F^* dual to F has been used.) Similarly on applying $Q^+\beta\gamma_5\gamma_\lambda$ to eq. (2.6.3) we get at first

$$F^{\mu\nu} Q^+\beta\gamma_5\gamma_\lambda [\gamma_\mu,\gamma_\nu] Q \psi = 0, \tag{2.6.7}$$

With the help of the identity:

$$\frac{1}{2}\gamma_5\gamma_\lambda[\gamma_\mu,\gamma_\nu] = - \varepsilon_{\lambda\mu\nu\rho} \gamma^\rho + \gamma_5 (g_{\lambda\mu}\gamma_\nu - g_{\lambda\nu}\gamma_\mu), \tag{2.6.8}$$

eq. (2.6.7) simplifies to:

$$F^{*\mu\nu} V_\nu\psi = 0. \tag{2.6.9}$$

It is the existence, in the presence of a nonzero external field, *of the two constraints* (2.6.6) and (2.6.9) *on* ψ, *that makes the system of equations* (2.6.2) *mutually inconsistent.* As a simple example, if we consider a constant magnetic field in the z-direction, these constraints become:

$$S_{03} \psi = V_o\psi = V_3\psi = 0. \tag{2.6.10}$$

The positive definiteness of V_0 then forces ψ to vanish identically.

A possible solution of this problem, which does not change the spectrum of solutions of the new Dirac equation, has been suggested by N. Mukunda, E.C.G. Sudarshan and C.C. Chiang [MUK 1]. It involves the replacement: internal bosons → parabosons. However, the possibility of a smooth classical limit seems thereby lost/ We pursue here a different route: we retain the freedom to rely on classical limiting forms, and allow the spin and mass to vary in a correlated way.

Finally, to be complete let us note that Staunton has found an elegant equation to describe the spin 1/2 states of the structure given in this chapter [STA 3]. This equation does not seem to allow interactions.

CHAPTER THREE

UNITARY REPRESENTATIONS OF THE POINCARE GROUP IN THE THOMAS FORM: QUASI-NEWTONIAN COORDINATES

§1. OVERVIEW

In order to obtain a deeper understanding of the nature of Dirac's new equation, (and especially its generalization to a Regge sequence) it is helpful to re-consider the famous construction (by Wigner [WIG 1]) of the unitary irreps of the Poincaré group, and to examine precisely how Dirac's new solutions fit in with this construction. This material is, to be sure, quite familiar, but our aim (and techniques) are somewhat novel, and this resurvey will prove rewarding. After first setting up the Wigner solutions (in §2), we will determine (in §3) the momentum space operators which generate the Wigner irreps. Next we resolve (in §4) the question as to how to determine the proper configuration space variables for the Wigner irreps and in this way recover the Poincare generators in the form* first found by L. H. Thomas [THO 1,2,3]. These configuration space variables have unusual properties (which we discuss); they are in fact *quasi-Newtonian coordinates* (in which $\vec{X},t$) do not form a four vector), and for these coordinates electromagnetic interactions are known to be impossible in general.

Quasi-Newtonian coordinates turn out to have one major advantage: using these coordinates it is easy to generalize the structure of the Thomas generators so as to obtain the mass-spin relation of a Regge band. Such a construction is one of the major goals of these lectures and this construction is discussed in §5.

The existence of a Poincaré covariant model which incorporates both mass and spin mixing (the Regge band discussed in §5) is possibly surprising, especially if one recalls the well-known series of "no-go" theorems (MacGlinn [MAC 1], O'Raifeartaigh [ORA 1], Segal [SEG 1] designed to rule out the existence of such models. Theorems have hypotheses and to circumvent an unwanted conclusion of a theorem one need only avoid one or more of the hypotheses. The standard way [WES 1], [HAA 1] to avoid O'Raifeartaigh's theorem is to adjoin Grassmann elements effecting Bose-Fermi transformations, and it is widely believed this is the only way. The construction of §5 is, however, another way [VAN 2,3], and it is the merit of the Thomas form (and quasi-Newtonian

---

* See also Shirokov [SHI 1] and Foldy [FOL 1] who developed similar results.

coordinates) to implement this construction in an elementary way. In §6 we extend this construction to yield a non-trivial, fully covariant supersymmetry.

Quasi-Newtonian coordinates were introduced into the usual Dirac electron equation in an elegent way by Foldy and Wouthuysen [FOL 1], in a procedure now widely known as the "Foldy-Wouthuysen transformation." Such transformations are by no means confined just to the usual Dirac electron equation, and we develop in a subsequent chapter (Chapter 4) the appropriate "inverse F-W transformation" to take the Thomas form solutions of the present chapter into the Minkowski coordinates of Chapter 4. This, as we shall show, is the necessary first step to obtain electromagnetic interactions for systems characterized by a Regge trajectory.

§2. THE WIGNER IRREPS [M,s]

We use the name Poincaré group, denoted by P, for the group of proper orthochronous inhomogeneous Lorentz transformations. The elements of the covering group of P are (d,A), where d represents a space-time translation and A a complex 2 x 2 matrix with unit determinant. Invariance for the group P leads, for quantum mechanics, to a unitary representation of the covering group of P.

The unitary irreducible representation (unirrep) which corresponds to a particle of spin 1/2, mass M and positive energy represents the Poincare element (d,A) by the transformation

$$[U(d,A)\phi]_i(p) = e^{ipd} \sum_{j=1}^{2} (B_p^{-1} A B_{p'})_{ij} \phi_j(p') ,$$

where

$$p' \equiv \Lambda^{-1}(A)p, \quad p^2 = p'^2 = M^2 , \quad p^o > 0, \ i = 1,2. \tag{3.2.1}$$

Here B_p stands for the boost (pure Lorentz transformation in the plane of the time axis and the four-vector p) which transforms the (reference) four-vector (0,0,0,M) into p. The hyperbolic angle ϕ of this boost is found from $p_o = M \cosh \phi$. The transformation B_p is then found to be

$$B_p \equiv \exp(\tfrac{1}{2}\phi\vec{\sigma}.\hat{p}), \tag{3.2.2}$$

where $\hat{p}$ is the unit three-vector in the direction of the three-vector $\vec{p}$ and $\vec{\sigma}$ stands for the Pauli matrices. Equation (3.2.2) can be written as

$$B_p = (\{p/M\})^{1/2}, \tag{3.2.3}$$

where

$$\{p/M\} = \frac{1}{M}\{p_o \underline{1} + \vec{p}\cdot\vec{\sigma}\} . \tag{3.2.4}$$

Note that

$$(\{p/M\})^{-1} = \frac{1}{M}\{p_o \underline{1} - \vec{p}\cdot\vec{\sigma}\} . \tag{3.2.5}$$

By expressing $\cosh\frac{1}{2}\phi$ in terms of $\cosh\phi$, one can rewrite (3.2.3) in the form

$$B_p = \frac{(M+p_o)\underline{1}+\vec{\sigma}\cdot\vec{p}}{[2M(M+p_o]^{\frac{1}{2}}} . \tag{3.2.6}$$

From (3.2.2) and (3.2.6) it is clear that

$$B_p^{-1} = \frac{(M+p_o)\underline{1}-\vec{\sigma}\cdot\vec{p}}{[2M(M+p_o)]^{\frac{1}{2}}} . \tag{3.2.7}$$

Returning to eq. (3.2.1), the functions $\phi_i(p)$ clearly satisfy

$$(p^2 - M^2)\phi_i(p) = 0. \tag{3.2.8}$$

Notice that the functions $\phi_i(p)$ are accordingly really functions only of the three-vector $\vec{p}$.

The Lorentz transformation corresponding to $B_{p'}^{-1}AB_p$ leaves the time axis invariant [since B_p changes (0,0,0,M) into p, A changes p into $\Lambda(A)p = p'$, and $B_{p'}^{-1}$ brings p' back to (0,0,0,M)] and is therefore a rotation of the little group. Hence the 2 x 2 matrix, $B_{p'}^{-1}AB_{p'}$, is *unitary*.

The invariant inner product of two wave functions is given by

$$(\phi,\psi) \equiv \int \frac{d\vec{p}}{p_o} \sum_{i=1}^{2} \phi_i^*(\vec{p})\psi_i(\vec{p}) . \tag{3.2.9}$$

For a unitary matrix A (which hence corresponds to a rotation), one finds from the definition of B_p, or via calculation using (3.2.6), that

$$B_{p'}^{-1}AB_p = A \text{ for } A \in SU(2). \tag{3.2.10}$$

The case of particles of spin s and mass M is entirely analoguous to the spin-½ case. The only change is that the 2 x 2 unitary matrix $B_{p'}^{-1}AB_p$ is replaced by its (2s + 1)-dimensional unitary representation $D^{(s)}_{mm'}(B_{p'}^{-1}AB)_p)$, and that ϕ now has 2s + 1 components. For the case of spin s, eq. (3.2.1) reads (cf. [FOL 1], [SHI 1]):

$$[U(d,A)\phi]_m(p) = e^{ipd} \sum_{m'=-s}^{s} D^{(s)}_{mm'}(B_{p'}^{-1}AB_p)\,\phi_{m'}(\Lambda^{-1}(p)). \tag{3.2.1'}$$

The functions $\phi_m(p)$ satisfy eq. (3.2.8), but (3.2.9) has to be replaced by

$$(\phi,\psi) = \int \frac{dp}{p^o} \sum_{m=-s}^{s} \phi^*_m(\vec{p})\psi_m(\vec{p}). \tag{3.2.9'}$$

Once again for $A \in SU(2)$ one has the special result

$$D^{(s)}_{mm'}(B_{p'}^{-1}AB_p) = D^{(s)}_{mm'}(A), \quad m=-s, -s+1,\ldots,s. \tag{3.2.10}$$

§3. POINCARÉ GENERATORS FOR THE WIGNER IRREPS

Having constructed the set of $\{M,s\}$ $\mathcal{P}$ irreps, it is important next to construct the explicit operators (observables) generating the symmetry structure. As is well known, symmetry considerations play a dual role in quantum physics, and lead not only to the structure of allowed states but also imply the proper observables (operators) of the symmetry.

It is clear from eq. (3.2.1) that the displacement generators, $\{P_\alpha\} = (P_0,\vec{P})$, take the eigenvalues

$$\vec{P} \to \vec{p}, \tag{3.3.1}$$

$$P_o = (\vec{P}^2 + M^2)^{1/2} \to +(\vec{p}^2 + M^2)^{1/2}. \tag{3.3.2}$$

Noting that the three-space rotations obey the simplifying result, eq. (3.2.10), it is easily verified that the rotation generators are [(ijk) = positive permutation of (123)]

$$M_{ij} = ip_i \frac{\partial}{\partial p_j} - ip_j \frac{\partial}{\partial p_i} + \sigma_k/2. \tag{3.3.3}$$

The really interesting generators are the boosts, M_{0i}, and it is somewhat more difficult to verify [from eq. (3.2.1)] that these have the form ([THO 1,2,3], [FOL 2], [SHI 1]:

$$M_{0i} = (\vec{p}^2 + M^2)^{1/2} \frac{\partial}{\partial p_i} + \frac{p_j\sigma_k - p_k\sigma_j}{2[(\vec{p}^2 + M^2)^{1/2} + M} . \tag{3.3.4}$$

To generalize from spin ½ to spin j one need only replace the 2 x 2 matrices $\sigma_i/2$ by the corresponding (2j + 1) x (2j + 1)- spin matrices $\vec{S}$ in eqs. (3.3.3) and (3.3.4).

Note that--for the inner product given by eq. (3.2.9)--the operator M_{0i} in eq. (3.3.4) is indeed Hermitian.

It is readily verified, directly, that the ten generators given in eqs. (3.3.1) - (3.3.4) close upon the commutation relations of the Poincaré group. [This uses the commutation rules $\vec{S} \times \vec{S} = i\vec{S}$, where $\{\vec{S}\}$ are the $(2j + 1 \times 2j + 1)$ matrix realizations of the generators of SU(2).] Note that time displacements are generated by the Hamiltonian $H = P_0 = +(\vec{P}^2 + M^2)^{1/2}$, $P^0 > 0$; hence the time, t, cannot correspond to an operator but functions, correctly, as a c number.

It is clear that the Wigner irreps, $\{M,s\}$, directly imply the momentum-space operator realizations given above; nonetheless these operators are of importance on their own. These operators have been rediscovered many times, in addition to [THO 1,2,3],[FOL 1],[BEC 1] by Bacry [BAC 1]. (Newton and Wigner [NEW 1] gave the results only for spin zero.)

§4. QUASI-NEWTONIAN COORDINATES AND THE GENERATORS IN THOMAS FORM

The Wigner irreps $\{M,s\}$ and the Wigner form of the generators are realized in momentum space, and it is an interesting question as to how to obtain configuration-space realizations. This problem was actually considered prior to the Wigner construction (1939) by Schrödinger, in his studies on the Dirac equation. As we shall show here, a comprehensive view can best be obtained directly from the Wigner construction of the set $\{M,s\}$; from this point of view the existence of the Dirac equation (for spin ½) is irrelevant, and the historical accident that this equation came first has greatly confused the initial, and subsequent, discussions of the problem, based as they are on the particularities of the Dirac equation itself.

The problem is this: How shall we introduce configuration-space variables of the Wigner irreps? At first glance, the question appears trivial; one should simply use a Fourier transformation. The difficulty is that the Wigner irreps are defined only on the mass hyperboloid in four-dimensional momentum space, and this constraint implies that the concept "Fourier transformation" is ill defined.

There are two distinct ways to proceed, leading to two very different results. We will designate the coordinates defined by these procedures as follows:

(a) Minkowski coordinates $\{x_\mu\}$ (which turn out to be the coordinates appropriate for coupling to the electromagnetic field), and

(b) Quasi-Newtonian coordinates $(\vec{X},t)$, which have been discussed in the literature by a great many authors.

To introduce the quasi-Newtonian coordinates one *postulates* the Fourier transform to be

$$\phi_i(\vec{X},t) = (2\pi)^{-3/2} \int d\vec{p}(p_0)^{-1/2} e^{i\vec{p}\cdot\vec{X}} e^{ip_0 t} \phi_i(\vec{p}). \tag{3.4.1}$$

From eqs. (3.2.9), or (3.2.9'), one finds that the norm (ϕ, ψ) may be written in terms of these new functions as

$$(\phi,\psi) = \int d\vec{X} \sum_i \phi_i^*(\vec{X},t)\psi_i(\vec{X},t). \tag{3.4.2}$$

Once having this definition, the form taken by the Poincaré generators can be found directly from the Wigner realization, eqs. (3.3.2)-(3.3.4):

$$P_i = -i\partial/\partial X_i, \tag{3.4.3}$$

$$\vec{M} = \vec{X} \times \vec{P} + \vec{S}. \tag{3.4.4}$$

From eq. (3.3.2) one finds the Hamiltonian to be

$$P_0 = +(\vec{P}^2 + M^2)^{1/2}. \tag{3.4.5}$$

Finally one finds for the generators of the Lorentz boosts [using eq. (3.3.4), and by partial integration in eq. (3.4.1).]

$$\{M_{0i}\} \equiv \vec{K} = \tfrac{1}{2}(\vec{X}P_0 + P_0\vec{X}) + tP_0^{-1}\vec{P} - (P_0 + M)^{-1}\vec{S} \times \vec{P}. \tag{3.4.6}$$

Several important remarks are to be made at this point.

(1) The generators (3.4.3), (3.4.4), (3.4.5) and (3.4.6) are Hermitian for the inner product (3.4.2). That they satisfy the commutation rules for the Poincaré group is clear by the way they were derived from the Wigner generators; one can also verify this fact directly. Note that t is once again a c number.

(2) Keeping (3.4.3), (3.4.4), and (3.4.5) one can modify (3.4.6) so that it remains Hermitian and keeps the correct Poincaré commutation relations. The most general form is

$$\vec{K}' = \vec{K} + f(p_0)\vec{P}, \tag{3.4.6'}$$

with $f(p_0)$ real. This freedom corresponds to the fact that instead of (3.4.1) one might have chosen

$$\phi_i(\vec{X},t) = (2\pi)^{-3/2} \int d\vec{p} e^{ig(p^0)} (p_0)^{-1/2} \times e^{i\vec{p}\cdot\vec{X}} e^{ip_0 t} \phi_i(\vec{p}). \tag{3.4.1'}$$

In other words, these representations, corresponding to different choices $f(p^0)$, are all unitarily equivalent.

(3) For the case of spin s the only change is that the Pauli matrices of the spin-½ case are replaced by the generators of the $(2s + 1)$ dimensional unitary representation of the quantal angular momentum group SU(2). Hence we have obtained quasi-Newtonian coordinates for the general (M,s) representation uniformly.

(4) The coordinates $(\vec{X},t)$ have a number of less desirable properties; this has been discussed in the literature extensively

(5) To our knowledge eqs. (3.4.3)-(3.4.6) were first given in this generality by Thomas [THO 1,2,3].

The coordinates $(\vec{X},t)$ have been designated here as "quasi-Newtonian" since $\vec{X}$ transforms under rotations as a three-vector, *but* $\vec{X}$ *is not part of a* (Minkowski) *four-vector*· as mentioned before, t is a c-number. These coordinates are therefore nonrelativistic in appearance, i.e., "Newtonian"; but since $\vec{X}$ properly belongs to quantum mechanics as an operator (or q-number) we accordingly call these coordinates quasi-Newtonian for short.

Such coordinates have been introduced many times but it is chiefly the discussions of Newton and Wigner [NEW 1]--"Newton-Wigner position operator"--and of Foldy and Wouthuysen [FOL 1]--"mean position operator"-- that have been definitive. There are major problems posed by quasi-Newtonian coordinates (localization is not invariant to Lorentz transformations; moreover, in the next instant the system is completely dislocalized). Physically the difficulty is that no interaction couples to this coordinate.

We defer the discussion of the Minkowski position coordinates to Chapter 4.

§5. GENERALIZATION OF THE THOMAS FORM

The proof that the Thomas form generators satisfy the correct Poincaré group commutation relations is by direct verification. All one needs for this verification are the two properties: (1) that the generators $\vec{S}$ have the commutation relations of angular momentum and (2) that these generators commute with $\vec{P}$ and with M. (For unitarity the three generators $\vec{S}$ need to be self adjoint.) This suggests the following generalization of Thomas's construction:

(a) Let H_{orb} be a Hilbert space of square integrable functions of $\vec{p}$,

(b) Let H_{int} be a Hilbert space which contains a unitary representation of SU(2) (this representation need not be irreducible) with generators $\vec{S}$, and which contains a positive, self-adjoint, operator M^2 which commutes with $\vec{S}$, that is, we have

$$[S_i , S_j] = i\varepsilon_{ijk}S_k , \tag{3.5.1}$$

$$[M^2, \vec{S}] = 0. \tag{3.5.2}$$

As M^2 is positive it has a unique positive square root M and, moreover, M commutes with S.

One can then easily verify the following result:

Lemma: There exists on $H_{int} \times H_{orb}$ a unitary representation of the Poincare group (in general, reducible) generated by the operators:

$$P = -i\partial/\partial X_i , \tag{3.5.3}$$

$$P_0 = (P^2 + M^2)^{1/2} \tag{3.5.4}$$

$$\vec{M} = \vec{X} \times \vec{P} + \vec{S} \tag{3.5.5}$$

$$\vec{K} = \frac{1}{2}(\vec{X}P_0 + P_o\vec{X}) + t P_0^{-1}P + (P_0 + M)^{-1} \vec{S} \times \vec{P}. \tag{3.5.6}$$

Any group G_j which acts on H_{int} will be called an *internal dynamical group*. In the next chapter we shall see several explicit examples of this general construction.

Remark: This lemma shows that relativistic SU(6) is clearly possible as there is no problem in constructing H_{int} so that it contains the generators of SU(6) as an extension of the three generators $\vec{S}$. We know $\vec{S}$ does not change $\vec{S}^2$ (the spin) or M^2, but this will not be necessarily true for the other generators of SU(6). [It is useful to note that this result corroborates a suggestion made by Gürsey [GUR 1] some time ago: that the introduction of relativistic SU(6) should be done in Foldy-Wouthuysen coordinates; a similar remark was made by Sudarshan and Mahanthappa [SUD 1].

§6. APPLICATION OF THE GENERALIZED THOMAS FORM: REGGE TRAJECTORIES

It is relatively easy now to apply the method of lemma of §5 to construct a model of a composite particle whose Poincaré invariant discrete labels of mass and spin are constrained to lie on a given Regge trajectory, $M^2 = f(s)$. To do so, we first recall that in the rest frame the spin zero state of the new Dirac equation, as well as the higher spin states, admitted the three spin generators $\vec{S}$ constructed from 2 bosons, a_1 and a_2. If we simply take the mass to be a function of the number of boson quanta--that is (to within a constant), a function of the SO(3,2) generator

$$V_0 = \frac{1}{2}(a_1^+a_1 + a_2^+a_2 + 1)$$

then we satisfy the conditions of the lemma. That is, we take:

$$M^2 = f(V_0), \tag{3.6.1}$$

and recognize (from eq. (2.1.17) or directly) that:

$$[\vec{S},M^2] = 0. \tag{3.6.2}$$

To be completely explicit, we then define a wave function $\psi(\vec{p},\xi_1,\xi_2)$, with inner product:

$$(\psi,\chi) \equiv \int d^3p \int d\xi_1 \int d\xi_2 \quad \psi^*(\vec{p}_1,\xi_1,\xi_2)\,\chi\,(\vec{p}_1,\xi_1,\xi_2), \tag{3.6.3}$$

for which the (Hermitian) generators of the Poincaré group are:

$$\vec{P} \to \vec{p}, \tag{3.6.4}$$

$$P_0 = [\vec{P}^2 + (f(V_0))^2]^{1/2}, \tag{3.6.5}$$

$$\vec{M} = \vec{X} \times \vec{P} + \vec{S}, \tag{3.6.6}$$

$$\vec{K} = \frac{1}{2}(\vec{X}P_0 + P_0\vec{X}) + \frac{\vec{P} \times \vec{S}}{P_0 + f(V_0)}\ , \tag{3.6.7}$$

[Here $\vec{X} \equiv i\,\partial/\partial\,\vec{p}$, with V_0 and $\vec{S}$ given by (2.2.17).]

The generators given by (3.6.4)-(3.6.7) define a reducible unitary representation of the Poincaré group which contains each value of spin (half-odd integer as well as integer) *once and only once*.

The function $f(V_0)$--which for physical reasons should be positive definite--gives the mass for each value of the spin. In other words, the structure defined by eqs. (3.6.2)-(3.6.7) is precisely what is denoted in the literature as *a single Regge trajectory* $M^2 = f(s)$ *which includes each spin once*. (The proof of these statements is immediate and requires no discussion.)

The fact that the generators S and V_0 which enter into this construction form four of the ten generators of SO(3,2), as discussed in Chapter 2, §2, is a strong hint that there should be some sort of transformation (in which the remaining six generators play a role) taking the entire structure defined by eqs. (3.6.2)-(3.6.7) from Quasi-Newtonian coordinates into Minkowski coordinates. In other words there should exist an "inverse Foldy-Wouthuysen" transformation defined on the whole set of mass-spin states of the entire Regge trajectory. The existence of such a transformation is the subject of Chapter 4, where we shall see that, via this transformation, we can give the structure defined by eqs. (3.6.2)-(3.6.7) a manifestly Poincaré covariant form.

§7. SUPERSYMMETRY: RELATIVISTIC SU(6)

Returning to (3.64)-(3.67) we can easily adjoin to the generators the creation and annihilation operators a_i, a_i^+, $i = 1,2$. These added generators raise the total spin by one-half unit (for creation) or (for destruction) lower the spin by one-half unit; at the same time they change the mass to that of the next state in the Regge sequence. The generators a_i^+ thus allow us to go up (in spin) along the ladder while the a_i allow us to go down. The wave function $\psi(p,\xi_1,\xi_2)$ thus describes a "composite" particle which can change its spin by half-integral steps. This is not quite global supersymmetry as defined in terms of field theory, but we shall see in Chapter 4 that the usual global supersymmetry operators can however be constructed nonetheless starting from a_i and a_i^+ in a modified way.

A further extension of the structure defined by eqs. (3.6.2) - (3.6.7) gives a non-empty relativistic formulation of SU(6). All one needs to do is consider *three pairs*--labelled by I, II, III,-- of harmonic oscillators each having a set of spin generators given by eq. (2.1.17). Next in eqs. (3.6.2) - (3.6.7) replace $\vec{S}$ by the sum of the three $\vec{S}$'s:

$$\vec{S} = \vec{S}_I + \vec{S}_{II} + \vec{S}_{III}, \tag{3.7.1}$$

and similarly replace V_0 by the sum of the three V_0's:

$$V_0 = V_{0I} + V_{0II} + V_{0III} \,. \tag{3.7.2}$$

Then the mass will be given by the total number of quanta of a degenerate six-dimensional harmonic oscillator, that is, there will be SU(6) degeneracy. The total spin, however, is given by (3.7.1), so that, from angular momentum addition, a single SU(6) mass multiplet will contain therefore different values of spin. Let the total number of quanta be denoted by n; then the mass is given by $M = f(n+3)$. For each mass the associated SU(6) multiplet belongs to the (totally symmetric) SU(6) irrep [n0̇]. For example, by the standard reduction, the n = 0 multiplet is a single spin zero state; the n = 1 multiplet is a triplet of spin 1/2 states; the n = 2 multiplet contains a six-plet with spin 1 and an anti-triplet with spin 0; the n = 3 multiplet is the familiar SU(6) 56-plet.

The SU(6) structure so obtained is nontrivial; it is *not* a direct product of the Poincaré group with something else. This is because the SU(6) generators other than S change the spin of the states within the given SU(6) mass multiplet. O'Raifeartaigh's theorem does *not* apply to this model; we shall investigate the reasons why this assertion is valid at the end of the next chapter, after first giving a manifestly Lorentz covariant form to our model.

CHAPTER FOUR

EXPLICITLY POINCARÉ INVARIANT FORMULATION, RELATION TO SUPERSYMMETRY, NO-GO THEOREMS

In Chapter 3 we achieved a (highly reducible) unitary representation of the Poincare group, describing the quantum mechanics of an entire Regge sequence, each value of spin occurring once. This structure, we showed, allowed operators which increase and decrease the spin by 1/2 at the same time as changing the mass (supersymmetry); moreover this general structure allows the combination of Poincaré invariance and internal symmetry into a single (non-direct) product structure (relativistic SU(6)). Although invariance for the Poincaré group is ensured, the invariance is not "manifest" in the sense that the formulation is not expressed in terms of standard (Lorentz) vector fields, spinor fields, and the like. In this chapter we achieve a manifestly Poincaré invariant form realized in Minkowski space. We also discuss the relation to global supersymmetry and to the no-go theorems (of the O'Raifeartaigh type) by exploiting the structure of this Minkowski space re-formulation [VAN 2,3,4,5].

§1. THE TRANSFORMATION FROM QUASI-NEWTONIAN TO MINKOWSKI COORDINATES

Generalities. In the application of the generalized Thomas form of the generators (chap. 3, §6) we were successful in obtaining a Poincare covariant structure which obeyed the mass-spin relation of a Regge trajectory. This structure has many satisfactory features from a theoretical point of view, the major advantage simply being the fact that the structure *indubitably exists*, and demonstrates thereby the consistency of the theoretical concepts. Moreover, it is equally clear that the states of this 'composite object' are *all time-like* and physically acceptable.

But there are difficulties, and the structure must be improved. Foremost among these difficulties is that the system--although it carries denumerably many unitary irreps of the Poincaré group--does not realize this symmetry in any obvious or manifestly covariant way. This difficulty is intimately connected to a second basic flaw, that the coordinates (X,t) are Quasi-Newtonian, fail to form a four-vector, and are inherently incapable of (electro-magnetic) interactions (at least in a gauge-invariant minimal-coupling way).

All of these difficulties exist in the case of the (massive) spin-½ Wigner irrep, but, of course, this troubled no one owing to the historical accident that the famous Dirac electron equation--which resolved the difficulties by using Minkowski coordinates from the very beginning--was obtained over a decade *before* Wigner's work. It is a useful exercise, however, to ignore history and ask how--starting from the Wigner $\{m,\frac{1}{2}\}$ irrep--one could achieve the Dirac equation. This is the problem which Thomas solved [THO 3], a problem which was not explicitly clarified in the earlier work of Foldy and Wouthuysen [FOL 1].

Thomas showed that if one doubled the space of the Wigner $(m,\frac{1}{2})$ irrep, introducing the diagonal operator ρ_3 in this doubled space, then the (Thomas form) generators took the form:

$$P_0 \rightarrow \rho_3 \, [\vec{P}^2 + m^2]^{\frac{1}{2}} \tag{4.1.1a}$$

$$\vec{K} \rightarrow \rho_3 \, (P_0\vec{X} + \frac{\vec{\sigma} \times \vec{P}}{2(P_0 + m)}) \tag{4.1.1b}$$

$$\vec{P} \rightarrow -i \frac{\partial}{\partial X} \tag{4.1.1c}$$

$$\vec{J} = \vec{X} \times \vec{P} + \tfrac{1}{2}\sigma \tag{4.1.1d}$$

Thomas then showed that the Foldy-Wouthuysen transformation of the above operators* then yielded the standard Poincaré generators of the Dirac electron equation. That is,

$$P_0 \rightarrow H_{Dirac}, \quad \vec{P} \rightarrow -i \frac{\partial}{\partial x} \tag{4.1.2a,b}$$

$$\vec{K} \rightarrow H_{Dirac} \, \vec{x} + \tfrac{1}{2}\rho_1 \vec{\sigma} \, , \quad \vec{J} \rightarrow \vec{x} \times \vec{P} + \tfrac{1}{2}\vec{\sigma} \tag{4.1.2c,d}$$

with $\vec{x}$ the space-part of a four-vector.

How can this example help us solve our problem? To begin with, we must get rid of one common inference from this example which is quite misleading, namely that the transformation from quasi-Newtonian to Minkowski coordinates (as exemplified by the Foldy-Wouthuysen transformation) is inherently connected with the existence of negative energies. A hint that this is wrong comes from the fact that in the new (positive energy) Dirac equation Minkowski coordinates actually do appear.

---

*To be precise, Thomas's work went in the other direction: Dirac form to Thomas form.

The proper analogy which will resolve the problem is to recognize that for the Dirac electron equation the F-W transformation diagonalizes the Hamilton in the 2 x 2 ρ-space, and that the analog to this is that the *four-component* operator Ω of the new Dirac equation reduces in the case of the Wigner representations to the *two-component* operator $\begin{pmatrix} a_1^+ \\ a_2^+ \end{pmatrix}$.

A rather different formulation of what turns out to be exactly the same idea is that the key to the desired transformation lies in the concept of the aligned bosons, which incorporates in the creation operators for aligned bosons both creation and destruction operators of the (fixed frame) bosons. (The analog to this in the Dirac electron case is the mixing of big and little components (that is, components in ρ-space) under boosts.)

§2. EXPLICIT CONSTRUCTION OF THE TRANSFORMATION

Let us consider now the entire set of Wigner irreps $\{M = f(s), s\}$ as united into a system defined by a single Regge trajectory*; the Thomas form generators for this structure were defined in eqs. (3.6.4-7). Equivalently we can use the (momentum space) Wigner form generators (eqs. (3.3.1-4)).

In either form, one would like to replace the explicit spin matrices that appear in the generators and treat all spins as a whole by going over to the spin operators as, say, differential operators on the (ξ_1, ξ_2) variables. Such a step immediately runs into trouble: the mass operator (and hence the operator P_0) cannot have a sharp value unless the spin magnitude is itself sharp. It follows that the Wigner irreps, which for the discrete labels (s, m_s) have a sharp formal four-momentum $(\vec{p}, p_0 = [\vec{p}^{\,2} + M^2(s)]^{\frac{1}{2}})$ lose this property when the continuous (ξ_1, ξ_2) labels are used.

On the other hand, it appears essential to introduce the labels (ξ_1, ξ_2) in the ket vectors, because in this way one can factorize the Wigner rotation into well-defined Lorentz boost operators.

The two sets of variables (ξ_1, ξ_2) and (s, m_s) are, however, completely equivalent so why cannot the Wigner rotation be factorized

---

*This system actually splits, via the univalence superselection rule, into two separate structures: integer vs half-integer spins, but this is not of consequence for the discussion at hand.

in *either* set of variables? The answer is that the spin variables (s,m_s) of the Wigner irreps *are not the correct variables to describe spins in the Minkowski frame*! The reason can be easily seen by example: if we were to factorize the Wigner rotation $B^{-1}_{p'}AB_p$ by using the SO(3,2) generators S_{AB} of eq. (2.2.17) to extend the spin S_{ij} to a Lorentz action, we would immediately find *that intermediate states having different spins* (s) *would be required*. This is physically unacceptable (since Lorentz transformations do *not* change the intrinsic spin), and clearly this shows that such a procedure uses an incorrect identification of the spin.

The correct spin variables to be used are the spin operators associated with the aligned bosons. These operators yield the same value of the intrinsic spin magnitude in every Lorentz frame. *Only in the rest frame do they agree with the spin value associated with the label* s *in the* {M(s),s} *Wigner irreps.*

This is the key observation and we can now develop the desired "inverse F-W" transformation directly from the Wigner irreps.

Consider the action of a boost A on the entire set of Wigner irreps {M(s),s}. This yields [Chap. 3, §2, using ket vectors instead of wave functions]:

$$U(0,A)\,|\,p(s);\ s,m_s\rangle \;=\; \sum_{m'_s} D^s_{m'_s m_s}(B^{-1}_{p'}\,AB_p)\,|\,p'(s);\ s,m'_s\rangle . \qquad (4.2.1)$$

(Here we have defined: $p(s) = (\vec{p}, p_0 = [\vec{p}^{\,2} + M^2(s)]^{\frac{1}{2}})$

and $p'(s) = \Lambda_A(p(s))$, where Λ_A is the Lorentz four-vector transformation associated to the Lorentz transformation A.)

Our goal is to suitably interpret the rotation matrix element so as to factorize it into well-defined Lorentz transformations associated with the Wigner rotation $B^{-1}_{p'}\,AB_p$.

The rotation matrix element is defined as a matrix element on the spin kets $|s,m_s\rangle$ only. If we seek to re-interpret these kets as kets for the Wigner irreps--that is, to adjoin the kets for the spatial part of the system--what momenta shall we use? The off-hand answer (use the initial and final momenta, p and p') is incorrect. The proper momenta to adjoin are the rest frame momenta. (This can be seen from the fact that the operators $B^{-1}_{p'}\,AB_p$ take the rest frame momentum $\underset{\circ}{p}$ first to p, then to $\Lambda_A(p) = p'$, and then back to $\underset{\circ}{p}$.) Thus the desired matrix element is:

$$\sum_{m'_s} D^s_{m'_s,m_s}(B^{-1}_{p'}\,AB_p)\,|\,\underset{\circ}{p}(s);\ sm'_s\rangle \;=\; U(B^{-1}_{p}AB_p)\,|\,\underset{\circ}{p}(s);sm_s\rangle \;. \qquad (4.2.2)$$

Now we make use of the facts that:

(a) Wigner kets in the rest frame are identical to "Minkowski kets" (that is, kets associated with the aligned boson operators) in the rest frame,

and (b) the operators B_p and A have a valid interpretation as unitary transformations generated by the operators $M_{\mu\nu} = L_{\mu\nu} + S_{\mu\nu}$, where $L_{\mu\nu}$ are the spatial Lorentz generators $i(P_\mu \frac{\partial}{\partial P\nu} - P_\nu \frac{\partial}{\partial P\mu})$ and $S_{\mu\nu}$ are the boson operators given by eqs. (2.2.17). (Note that this interpretation agrees with the fact that $B_{p'}^{-1} AB_p$ is a rotation generated by the spin operators of eq. (2.1.17) alone, since spatial rotations are trivial when acting on rest frame kets.)

It is essential at this point to clarify the notation so as to make the results to follow unambiguous in content. The transformations B_p and A in the Wigner development of Ch. 3, §2 originally denoted 2 x 2 (SL(2,C)) matrices representing the Lorentz boost $\underset{\circ}{p} \rightarrow p$ and a generic Lorentz transformation respectively. The product $B_{p'}^{-1} AB_p$ then denoted a unitary 2 x 2 (rotation) matrix, which then was interpreted in $D^s(B_{p'}^{-1} AB_p)$ as the 2s+1 x 2s+1 rotation matrix for the SU(2) element denoted by $B_{p'}^{-1} AB_p$. We are now going to reinterpret these 2 x 2 SL(2,C) matrices as unitary matrices whose generators are the Hermitian $M_{\mu\nu}$ operators. Thus to be precise, we will denote these re-interpreted matrix operators as $\mathcal{U}(B_p)$, $\mathcal{U}(A)$,..., so that $\mathcal{U}$ operators represent the Lorentz SL(2,C) elements of the Wigner construction. Let us note that $\mathcal{U}(B_{p'}^{-1} AB_p)$ is still a rotation, and in fact precisely the same rotation in its action on the $|\underset{\circ}{p}(s);\ sm_s>$ eigen-kets as the original Wigner rotation $B_{p'}^{-1} AB_p$. The important difference is that in this new form the Wigner rotation may be factored, so that $\mathcal{U}(B_{p'}^{-1} AB_p) = \mathcal{U}(B_{p'}^{-1})\ \mathcal{U}(A)\ \mathcal{U}(B_p)$. Before writing this result out for eq. (4.2.2) in detail, we observe that the action of $\mathcal{U}(B_p)$ on a rest frame ket yields an *aligned boson ket* (see Ch. 2, §4). That is:

$$\mathcal{U}(B_p)|\underset{\circ}{p}(s);\ sm_s> \ = |p(s);\ \ sm_s>_{\hat{p}} \qquad (4.2.3)$$

where subscript $\hat{p}$ on the ket denotes that the boson ground state $|0>$ is aligned with the unit four-vector p. It is useful to observe that this result is *independent* of the mass M(s), and that the spin (s,m_s) are *intrinsic* spin labels.

We can now factorize the matrix element in eq. (4.2.2). The result is:

$$\sum_{m'_s} D^s_{m'_s,m_s}(B^{-1}_{p'}AB_p)\ |\,p'(s);sm'_s>_{\hat{p}'} = \mathcal{U}(A)\ |\,p(s);sm_s>_{\hat{p}} \ . \qquad (4.2.4)$$

Before we can use this result in eq. (4.2.1), we must properly re-interpret the Wigner kets that appear in this equation. The Wigner ket $|p(s);\ sm_s>$ is the direct product of the spatial ket $|\vec{p},p_0 = [\vec{p}^2 + M^2(s)]^{1/2}>$ with the spin ket $|s,m_s>$ whose labels are defined by the action of the generators S_{ij} in eq. (2.1.17). To re-interpret this Wigner ket as a "Minkowski ket" with the intrinsic labels (s,m_s) we see that we must "align the bosons" by using the boost operator B_p, *in spin space only* (generators defined by eq. (2.2.17)). Thus we have:

$$|p(s);\ sm_s> = \mathcal{U}(B^{-1}_p)^{\text{spin}}\ |\,p(s);\ sm_s>_{\hat{p}} \ . \qquad (4.2.5)$$

Next we observe that a boost transformation is actually a function only of the three variables $\hat{p}$--and hence independent of the mass (=length of the four vector). Replacing the numerical parameters p by the *operator* $\hat{P} \equiv (P.P)^{-1/2}P$ we can define the operator $\mathcal{U}(B_{\hat{P}})$ on an arbitrary ket.

Using these results we now re-write equation (4.2.1) in the form:

$$U(0,A)\ \mathcal{U}(B^{-1}_{\hat{P}})^{\text{spin}}\ |\,p(s);\ sm_s>_{\hat{p}}$$

$$= \sum_{m'_s} \mathcal{U}(B^{-1}_{\hat{P}})^{\text{spin}}\ |\,p'(s);\ sm'_s>_{\hat{p}'}\ D^s_{m'_s,m_s}(B^{-1}_{p'}\ AB_p) . \qquad (4.2.6)$$

Operating on both sides of this equation with the operator $\mathcal{U}(B_p)^{\text{spin}}$, and then using eq. (4.2.4) yields the result:

$$\mathcal{U}(B_{\hat{P}})^{\text{spin}}\ U(0,A)\,\mathcal{U}(B^{-1}_{\hat{P}})^{\text{spin}}\ \ |p(s);\ sm_s>_{\hat{p}}$$

$$= \mathcal{U}(A)\ \ |p(s);\ sm_s>_{\hat{p}} \ . \qquad (4.2.7)$$

Since this result, eq. (4.2.7), is valid for arbitrary Minkowski (aligned) kets we can assert the operator identity:

$$\mathcal{U}(B_{\hat{P}})^{\text{spin}}\ \ U(0,A)\ \ \mathcal{U}(B^{-1}_{\hat{P}})^{\text{spin}} = \mathcal{U}(A). \qquad (4.2.8)$$

This is our basic result, which demonstrates *that the operator labelled spin boost,* $\mathcal{U}(B_{\hat{P}})^{spin}$*, is the desired "inverse F-W" transformation* which takes the unitary Lorentz transformations of the entire set of Wigner irreps $\{M(s),s\}$ into the unitary Lorentz transformations on the Minkowski space irreps. In other words, the operator $(B_{\hat{P}})^{spin}$ transforms the Quasi-Newtonian momentum coordinates into the Minkowski momentum coordinates, where the states in both sets of coordinates obey the mass restriction $P_0 = [\vec{P}^2 + M^2(s)]^{\frac{1}{2}}$.

It remains to determine the transformations for displacements, and for pure rotations. For rotations, the operator $\mathcal{U}(B_{\hat{P}})^{spin}$ commutes with the generators M_{ij} so that:

$$\mathcal{U}(B_{\hat{P}})^{spin}\ U(0,R)\ \mathcal{U}(B_{\hat{P}}^{-1})^{spin} = U(0,R) = \mathcal{U}(R), \qquad (4.2.9)$$

where $\mathcal{U}(R)$ is a rotation generated by the operator $M_{ij} = L_{ij} + S_{ij}$.

The situation for spatial displacements is similar to that for rotations, since the generators $\vec{P}$ commute with the operator-labelled spin boost $\mathcal{U}(B_{\hat{P}})^{spin}$. But for temporal displacements the situation is more complicated.

It is easiest to consider infinitesimal transformations, which--for the Wigner irreps--involves the generator: $P_0 = [\vec{P}^2 + M^2(s)]^{\frac{1}{2}}$. Here the spin s is the eigenvalue of $\vec{S}^2$, where $\vec{S}$ is the boson operator of eq. 2.1.17. *The operator,* $\vec{S}$, (and hence $M^2(s)$) *does not commute with* $\mathcal{U}(B_{\hat{P}})^{spin}$. Instead, under the action of $\mathcal{U}(B_{\hat{P}})^{spin}$, $\vec{S} \to \vec{S}(\hat{P})$. Thus we obtain the *intrinsic spin operators,* and accordingly $M^2(s)$ transforms into $M^2(\hat{P}) \equiv M^2_{op}$. Let us determine this operator explicitly.

First off one sees that M^2_{op}--since it is determined by the discrete eigenvalues of the intrinsic spin--is necessarily Poincaré invariant. The explicit result is not hard to obtain using the aligned bosons of Chapter 2, §4. One finds that:

$$M^2_{op} = \mathcal{U}(B_{\hat{P}})^{spin}\ (M^2(s))\,\mathcal{U}(B_{\hat{P}}^{-1})^{spin} = \alpha(P \cdot V), \qquad (4.2.10)$$

where $\hat{P} \equiv (P \cdot P)^{-\frac{1}{2}} P$, V is the boson operator four-vector defined in (2.1.17), and the function α is the Regge trajectory function.

In this form it is obvious that M^2_{op} is indeed Poincaré invariant.

The final form for the transformation from the set of all Wigner irreps $\{M(s),s\}$ to the Minkowski (momentum) space irreps is then given by:

$$\mathcal{U}(B_{\hat{P}})^{spin}\ U(d,A)\ \mathcal{U}(B_{\hat{P}}^{-1})^{spin} = e^{id\cdot P}\,\mathcal{U}(A) \quad , \qquad (4.2.11)$$

where $P = (\vec{P}, (\vec{P}^2 + M^2_{op})^{\frac{1}{2}})$, and Λ denotes any Lorentz transformation.

Summary. We have demonstrated in this section that there exists a transformation between the states of a composite system whose Regge trajectory is defined on the set of all Wigner irreps $\{M(s),s\}$ and the same composite system realized in Minkowski momentum space. This is the inverse F-W transformation which was sought.

This transformation transforms the generators of the Poincaré group in the two distinct realizations. Since the generators for the Wigner representations have already been given (Thomas form, eqs. 3.6.4-7); momentum space form eqs. (3.3.1-4)), it remains only to give these generators in Minkowski form.

We have already determined the displacement operators to have the form:

$$\vec{P} \rightarrow \vec{p}, \tag{4.2.12a}$$

$$P_0 = [\vec{P}^2 + M^2_{op}]^{\frac{1}{2}} , \tag{4.2.12b}$$

with
$$M^2_{op} \equiv \alpha(\hat{P} \cdot V)$$
$$= \alpha(V_0(B_{\hat{P}})).$$

(V_μ is defined in eq. (2.2.17)).

The Lorentz generators have the form:

$$M_{\mu\nu} = L_{\mu\nu} + S_{\mu\nu} \quad . \tag{4.2.12c}$$

where:

$$L_{\mu\nu} = P_\mu \frac{\partial}{\partial P\nu} - P_\nu \frac{\partial}{\partial P\mu}$$

and $S_{\mu\nu}$ is defined in (2.2.17).

The generators $L_{\mu\nu}$ are defined as if the operator P_0 were unrestricted, since the displacements generated are in any event tangent to the constraint surface. Thus one may equivalently consider the four-vector operator $P \rightarrow p$ to be sharp, and then restrict the states by the Poincaré-invariant condition:

$$P \cdot P = \alpha(\hat{P} \cdot V), \quad P_0 \rightarrow \text{positive eigenvalue}. \tag{4.2.13}$$

These results can be obtained by transforming eqs. (3.3.1-4) with the operator B_P^{spin}, but this direct approach is not as easily carried out for the boost operators as the 'global' techniques used above.

Remarks. (1) We have noted above that the rotation generators M_{ij}, commute with the transformation generated by $U(B_{\hat{p}})^{spin}$. However, the separate terms L_{ij} and S_{ij} of M_{ij} do *not* commute with this transformation. This leads to the important result that the spin operator S_{ij} and the transform of S_{ij} by $U(B_{\hat{p}})^{spin}$ are *distinct* operators. The latter operator is the "intrinsic spin" operator $S_{ij}(B_{\hat{p}})$.

(2) Unlike the spin and orbital angular momentum operators in the Thomas form--which separately are constants of the motion--the spin and orbital angular momentum operators in the Minkowski form are *not* separately conserved (do not commute with the Minkowski form operator P_0). The sum of the two operators (M_{ij}) is, of course, conserved.

This situation is closely analogous to the Foldy-Wouthuysen case for the electron equation, where the "mean spin" and "mean orbital angular" momentum operators (the Thomas form operators) are separately conserved.

(3) We have given only the momentum space form of the Minkowski generators. It is not difficult to give the configuration space form of these generators, since this can be found directly from the Fourier transform.

The displacement generators now take the form:

$$P_\mu = -i \frac{\partial}{\partial x_\mu} , \tag{4.2.14a}$$

and the Lorentz generators are:

$$M_{\mu\nu} = i\, x_\mu p_\nu - x\ \ p_\mu + S_{\mu\nu} , \tag{4.2.14b}$$

and the covariant wave-function constraint is formally as before:

$$P_0 = [\vec{P}^2 + \alpha(P \cdot V)]^{\frac{1}{2}} . \tag{4.2.14c}$$

(4) It is not difficult to write an arbitrary number of Wigner representations for different masses and spins into "one composite system". Such a structure is basically trivial and clearly ad hoc. It is important to discuss why the present composite structure escapes this criticism.

The reason is that the structure is not at all ad hoc but tightly constrained by the requirement that the representations are tied together as a unit by the fact that they realize--in any frame--the integer and half-integer irreps of the SO(3,2) group. The generators of this "kinematical symmetry group" are the SO(3,2) generators of eq.(2.1.17) transformed by $\mathcal{U}(B_{\hat{p}})^{spin}$ to the "aligned boson" realization. Clearly for every Lorentz frame these 10 transformed SO(3,2) generators realize the SO(3,2) symmetry.

In the rest frame, this symmetry is realized on the original set of Wigner representations, but in any other Lorentz frame we must use the Minkowski realizations developed above.

Alternatively one could demonstrate that the set of Wigner representations united into a whole is not trivial by demonstrating that there exist operators carrying one mass-spin representation into another. The required operators--valid for any Lorentz frame--would seem to be the aligned boson operators $a^{+}_{i}(\hat{P})$ and $a_{i}(\hat{P})$, but these are not quite sufficient. The reason is that although these operators correctly change the spin by $\pm$ 1/2 *they do not change the spatial wave function to the corresponding mass.* This defect is easily remedied by the formal introduction of a scaling operator, S, to be introduced in the next section. With the aid of this operator, we can define raising-lowering operators--valid in any Lorentz frame--that change the representation $(M(s),s)$ into $(M(s\pm\frac{1}{2}), s\pm\frac{1}{2})$. These operators are the operators: $Sa_{i}^{+}(\hat{P})S^{-1}$ for creation and $Sa_{i}(\hat{P})S^{-1}$ for destruction.

It is the existence of these operators that allows one to conclude that the composite structure defined by the Regge trajectory $M^2 = f(s)$ and the Minkowski space generators of eq. (4.2.14) is indeed a single entity tied together by operators linking adjacent mass-spin Poincaré irreps.

§3. AN ALGEBRA WHICH EXTENDS THE POINCARÉ ALGEBRA AND CONTAINS OPERATORS FOR RAISING AND LOWERING MASS AND SPIN

The generators $T_a(\hat{p})$ and $\overline{T}_a(\hat{p})$, discussed in Chapter 2,§4 change only the spin, leaving the mass unaffected, hence they do not leave our structure invariant. If we were to operate on the spin-zero state we obtain a spin-½ state with mass m_o instead of with the proper value $m_{\frac{1}{2}}$. To achieve the change of mass we need a scaling operator S defined by

$$[\mathcal{s},p] = ip, \ [\mathcal{s},x] = -ix \ . \qquad (4.3.1a,b)$$

Using $\mathcal{s}$ we define the (Lorentz invariant) scaling operator:

$$S = \exp[-i(\ell n M_{op})\mathcal{s}], \qquad (4.3.2)$$

such that:

$$S^{-1}PS = M_{op}^{-1}\,P\ , \tag{4.3.3}$$

(this uses $[P,M_{op}] = 0$).

The desired global four-component annihilation operator (lowering spin *and* changing mass) is then defined by

$$S_a = \tfrac{1}{2}(1 + i\gamma_\mu P^\mu M_{op}^{-1})_{ab}\ S^{-1}\ \Omega_b. \tag{4.3.4}$$

The conjugate raising operator S_a is then defined by:

$$\bar{S}_a = S_h^+(\gamma^o)_{ha}. \tag{4.3.5}$$

The operators S_a and $\bar{S}_a$ will be shown to transform as four-component spinors (not Majorana spinors). Moreover, these operators satisfy the relations:

$$[S_a,S_b] = 0 = [\bar{S}_a,\bar{S}_b] \tag{4.3.6}$$

$$[S_a,\bar{S}_b] = (-i)\ \tfrac{1}{2}(1 + i\gamma_\mu P^\mu M_{op}^{-1})_{ab}\ . \tag{4.3.7}$$

These algebraic relations are to be contrasted with the analogous (standard) supersymmetry algebra where commutation is replaced by anticommutation and where the mass term (the unit operator in the projection operator on the right hand side of (4.3.7) is missing. (We shall come back to this point; let us remind the reader here that so far we have not considered field theory but rather a relativistic wave equation for a particle which is able to have all values of spin.)

Note the identity: ($\{,\}$ denotes the anti-commutator).

$$\sum_a \{\bar{S}_a,S_a\} = 8\ P\cdot V/M_{op}. \tag{4.3.8}$$

The operators S_a (and $\bar{S}_a$) transform as spinors under the Lorentz group, that is,

$$[M_{\mu\nu},S_a] = (\sigma_{\mu\nu})_{ba}\ S_b, \tag{4.3.9}$$

$$[M_{\mu\nu},\bar{S}_a] = (\sigma_{\mu\nu})_{ab}\,\bar{S}_b\,, \tag{4.3.10}$$

where

$$\sigma_{\mu\nu} = \frac{1}{4}[\gamma_\mu,\gamma_\nu].$$

The most interesting commutation relations involve the translation generators P_μ where one finds:

$$[P_\mu,S_a] = (M_{op}S_aM_{op}^{-1} - S_a)P_\mu\,. \tag{4.3.11}$$

It is of great importance to note that whereas P did *not* commute with S_a, the "velocity" operator defined by PM_{op}^{-1} *does commute:*

$$[P_\mu M_{op}^{-1},\, S_a] = 0 = [P_\mu M_{op}^{-1},\, \bar{S}_a], \tag{4.3.12}$$

this shows clearly that the operators S_a, $\bar{S}_a$ lower and raise mass and spin keeping the four velocity $P_\mu M_{op}^{-1}$ invariant.

The expression (4.3.11) becomes particularly simple for a special case. Let us choose the trajectory relation to be linear having the form:

$$M_{op}^2 = M_o P\cdot V. \tag{4.3.13}$$

The choice (4.3.13) for M_{op} corresponds to a Regge sequence with mass proportional to spin, that is, the mass levels are spaced equidistantly with distance M_o along a given direction of the four velocity $P_\mu M_{op}^{-1}$. For that choice one has

$$[P_\mu,S_a] = -M_oS_aP_\mu M_{op}^{-1}\,, \tag{4.3.14a}$$

$$[P_\mu,\bar{S}_a] = M_o\bar{S}_aP_\mu M_{op}^{-1}\,. \tag{4.3.14b}$$

These are obvious relations, as on any eigenstate of P_μ the operations of first lowering the mass by M_o and then determining P_μ, and of first determining P_μ and then lowering the mass by M_o differ by $M_o\hat{P}_\mu = M_oP_\mu M_{op}^{-1}$.

Note that the relation (4.3.14a,b) contrast with the standard supersymmetry relation which has zero on the right hand side. However, if one seeks to unite multiplets of different mass, as we do here, then a relation of the form (4.3.11) must obtain.

In analogy to what is done in supersymmetry, one may regard the algebra generated by $\{P_\mu, M_{\mu\nu}, S_a, \bar{S}_a\}$, with commutation relations (4.3.6-7), (4.3.9-11) as an extension of the Poincaré algebra spanned by P_μ and $M_{\mu\nu}$. Notice, however that *this algebra does not close* and it is in this way that our construction avoids the no-go theorems. The nonclosure of the algebra is particularly simple for the special case (4.3.14a,b), where one generates terms $S_a P_\nu M_{op}^{-1} P M_{op}^{-1} \ldots$. As the algebra does not close there is no finite rank Lie group associated with the algebra. (The finiteness of the rank of the Lie state was the basic assumption made by O'Raifeartaigh and others in obtaining no-go theorems.)

Equation (4.3.12) is important in this connection because it shows clearly that the *algebra generated by* $\{P_\mu M_{op}^{-1}, M_{\mu\nu}, S_a, \bar{S}_a\}$ *does close*. In other words, the algebra which we are considering is well defined, in fact it is quite natural from a physical point of view. Consider a system in its rest frame that can exist in several discrete mass states. A boost transformation changes four-velocity so that the set of discrete mass states considered above will *all* have the same four-velocity. The possibility of forming coherent combinations of wave functions with differing mass and spin (subject to super-selection rules) thus exists in each Lorentz inertial frame indexed by a unit four-velocity.

The algebraic Lie group associated to the algebra of the preceding discussion defines an interesting physical structure. One may view this structure geometrically in terms of the three-dimensional surface of a unit-mass hyperboloid in the forward light cone. For every four-velocity, there is a corresponding point of the surface; over each point we may associate a set of different mass-spin states. The vector space of these states is precisely the fiber of a principle fiber bundle, whose base space consists of the points of the hyperbolic surface. The isotropy group of a given point (the transformations leaving a given four-velocity invariant) is isomorphic to the group SU2 extended by the shift operators S_a, $\bar{S}_a$, adapted to this velocity. This situation is illustrated in Fig. 4.1.

§4. RELATION WITH STANDARD GLOBAL SUPERSYMMETRY

The present model, with wave function $\phi(p,\xi_1,\xi_2)$, and wave equation (4.2.13c) describes a single particle which can have all values of spin, integer as well as half-integer. At any fixed value of $P^\mu M_{op}^{-1}$ the generator $\bar{S}_a$ raises the spin by 1/2 and changes the mass appropriately; S_a lowers mass and spin. In figure 4.2 we have indicated this explicitly for the special case of equidistant levels (4.3.14) with distance M_o.

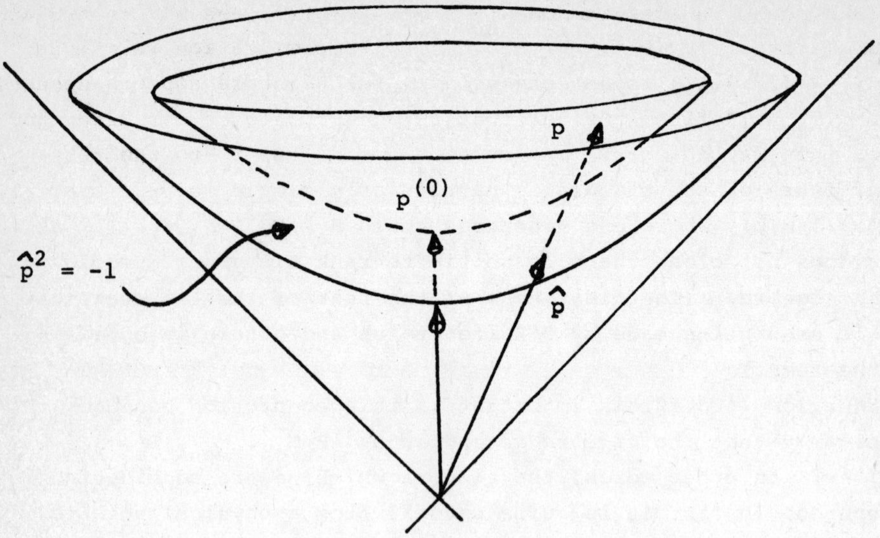

Figure 4.1

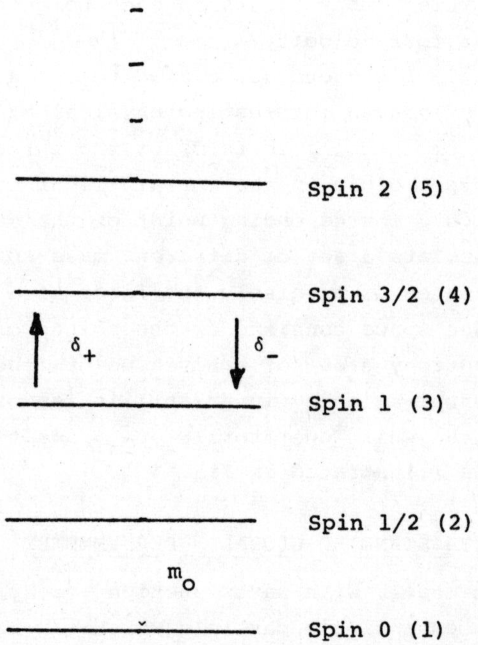

Figure 4.2

The variation $\delta_+ \phi(p,\xi_1,\xi_2)$ defined by:

$$\delta_+ \phi(p,\xi_1,\xi_2) = \eta_a \bar{S}_a \phi(p,\xi_1,\xi_2) , \qquad (4.4.1)$$

is indicated in figure, as is

$$\delta_- \phi(p,\xi_1,\xi_2) = \bar{\eta}_a S_a \phi(p,\xi_1,\xi_2) . \qquad (4.4.2)$$

The wave function $\phi(p;\xi_1,\xi_2)$ can be decomposed into a complete set of states indexed by the value of the spin. The key to this analysis into spin eigenfunctions is the introduction of creation and annihilation operators which parametrically contain the unit velocity vector $\hat{P} \equiv P/M$, as well as the introduction of the vacuum (ground state, spin zero) eigenket which must also parametrically depend on the four-velocity $\hat{P}$. This structure has already been achieved in connection with the new Dirac equation, and has been discussed under the concept of aligned bosons (Ch. 2, §4).

In the rest frame--denoted by $\overset{\circ}{p} = (000\ m)$--the ground state for the spin part of the new Dirac equation is:

$$u_0(\overset{\circ}{p}) \equiv \langle \xi_1 \xi_2 | 0,0 \rangle = e^{-\frac{1}{2}(\xi_1^2 + \xi_2^2)} . \qquad (4.4.3)$$

(Here $|0,0\rangle$ denotes the ket with $N_1 = 0$ and $N_2 = 0$ bosons.) The ground state corresponding to a general momentum $p = \Lambda(p)$, with $p \cdot p = -m_0^2$, is then obtained by boosting the rest frame solution using the operator B_p (which acts only on the spin variables):

$$u_0(p) \equiv \langle \xi_1 \xi_2 | B_p | 0,0 \rangle . \qquad (4.4.4)$$

The boost $\Lambda(p)$, which takes $\overset{\circ}{p}$ to p, and may be chosen to the Lorentz transformation in the plane spanned by $\overset{\circ}{p}$ and p. The very same technique may be used to obtain all the excited states at a general p from the excited states in the rest frame $\overset{\circ}{p}^{(0)}$. It is essential to observe that the boost $\Lambda(p)$ is precisely the same for all those p which have the same four direction (four velocity) $\hat{p}$, $p \cdot p = -1$, that is, the boost $\Lambda(p)$ is independent of the mass of the state transformed.

The creation-destruction operators for states having the unit four-velocity $\hat{p}$ are the aligned operators:

$$a_j^+(\hat{p}) \equiv B_p(a_j^+)B_p^{-1} \quad , \tag{4.4.5a}$$

$$a_j(\hat{p}) \equiv B_p(a_j)B_p^{-1} \quad . \tag{4.4.5b}$$

Although these aligned operators transform collectively as a four-component spinor;

$$\Omega(\hat{p}) = \begin{pmatrix} a_1^+(\hat{p}) \\ a_2^+(\hat{p}) \\ a_2(\hat{p}) \\ -a_1(\hat{p}) \end{pmatrix} , \tag{4.4.6}$$

it is desirable to separate, covariantly, the creation from the destruction aspects for these aligned operators.

This is the function of the operators $T_a(\hat{p})$ and $\overline{T}_a(\hat{p})$, introduced in eq. (2.4.5) and (2.4.7). That is, we use the four-spinors:

$$T_a(\hat{p}) = [\tfrac{1}{2}(\mathbb{1} + i\gamma\cdot\hat{P})\Omega]_a \ , \tag{4.4.7}$$

and:

$$\overline{T}_a(\hat{p}) = [(T(\hat{p}))^+\gamma_0]_a \quad . \tag{4.4.8}$$

It is reasonable to call $T_a(p)$ and $\overline{T}_a(p)$, the "covariant creation" and "covariant annihilation" operators, respectively, since the operators are covariant and have the described aspect in the rest frame.

Using these operators we can now analyze the general wave function $\phi(p;\xi_1\xi_2)$ into spin components. Thus, the spin-zero state of our particle is described by a scalar wave function $\psi_o(p)$, with $p^2 + m_0^2 = 0$, which is obtained from the general wave function $\phi(p;\xi_1,\xi_2)$ by projection:

$$\phi_o(p) = \int d\xi_1 \, d\xi_2 u_0^* \, (\hat{p};\xi_1,\xi_2) \; \phi(p;\xi_1,\xi_2) . \tag{4.4.9}$$

Similarly the four functions $\overline{T}_a(\hat{p})u_0(\hat{p};\xi_1,\xi_2)$ are the basis states for spin 1/2. Hence the spin -1/2 state of our particle may be described by a four-spinor function, $\phi_a(p)$, given by:

$$\phi_a(p) = \int d\xi_1 d\xi_2 \; (\overline{T}_a(\hat{p})u_0(\hat{p};\xi_1,\xi_2)) \; \phi(p;\xi_1,\xi_2), \; a=1...4, \tag{4.4.10}$$

where again $\phi(p;\xi_1,\xi_2)$ is the wave function for the entire system.

The functions $\phi_a(p)$ satisfy:

$$(P^2 + m_{1/2}{}^2)\ \phi_a(p) = 0 \qquad (4.4.11)$$

and also:

$$(1\!\!1 + i\hat{p}_\mu\gamma^\mu)_{ab}\phi_a(p) = 0\ . \qquad (4.4.12)$$

The spin-1 state of $\phi(p;\xi_1,\xi_2)$ is similarly constructed by projecting on $T_a\overline{T}_b u_0(\hat{p})$ and is given by an object with two symmetrized spinor indices. In this way one can project out the components of $\phi(p;\xi_1,\xi_2)$ corresponding to any fixed spin, obtaining symmetric multi-spinor wave functions $\phi_{ab\ldots}(p)$. The mass spectrum given by $M_{op} = f(V_o)$ for the wave functions $\phi(\vec{p};\xi_1,\xi_2)$ must be imposed on $\phi(p;\ \xi_1,\xi_2)$ as an invariant subsidiary condition:

$$(P^2 - f^2(\hat{p}^\mu V_\mu))\ \phi(p;\xi_1,\xi_2) = 0. \qquad (4.4.13)$$

Denoting these wave functions by: ϕ_0, ϕ_a, ϕ_{ab}, ϕ_{abc}, ..., (where there is always symmetry in the spinor indices) one may translate the variations δ_+ and δ_- onto these wave functions. One finds that $\delta_{+\eta}$ translates as the action:

$$\delta_{+\eta}\psi_o = 0; \quad \delta_{+\eta}\psi_a = 1/2\ (\frac{\gamma\cdot p}{m_{1/2}} - i)_{ab}\ \eta_b\psi_o(+)\ ;$$
$$(4.4.14)$$
$$\delta_+\psi_{ab} = 1/2\ [(\frac{\gamma\cdot p}{m_1} - i)_{ab}\eta_b\psi_\beta(+) + (\frac{\gamma\cdot p}{m_1} - i)_{bc}\eta_c\psi_a(+)]\ldots,$$

where the plus sign on the right-hand side means that the mass has been scaled up from m_j to $m_{j+1/2}$.

For the variation δ defined by $\overline{\eta}_\beta S_\beta$ one finds in a similar way:

$$\delta_{-\eta}\psi_o = \overline{\eta}_b\psi_b(-)\ ; \qquad \delta_-\psi_a = \overline{\eta}_b\psi_{ab}(-);\ \ldots \qquad (4.4.15)$$

Let us emphasize that the actions represented by eqs. (4.4.14) and (4.4.15) map solutions of eq. (4.4.13) into solutions; the parameters η are c-numbers at this stage.

Next we consider the *second-quantized form* of the solutions to eq. (4.4.13); this can be done by second-quantizing the wave functions $\phi_{ab\ldots}$ which now become free field operators. The spin-statistics theorem implies that field operators with an odd number of Dirac indices anti-commute; this requires that the parameters η in eqs. (4.4.14) and (4.4.15) must be Grassmann variables anti-commuting with themselves and all odd-indexed field operators.

Using the operators $\delta_{+\eta}$ and $\delta_{-\eta}$, which for field theory (because of η being a Grassmann variable) anti-commute, and which together connect all the levels of the system, one can construct "supersymmetry operators" which cannect only a *finite* number of levels. For instance one can define an operation δ_η on the levels of spin 0 and spin 1/2 by defining δ_η to be $\delta_{+\eta}$ when acting on the spin zero state and to be $\delta_{-\eta}$ when acting on the spin 1/2 states. One finds with (4.4.14) and (4.4.15):

$$[\delta_{\eta'},\delta_\eta]\psi_o = -2(\eta'^T\gamma^o\gamma^\mu\eta)P_\mu\psi_o \tag{4.4.16}$$

$$[\delta_{\eta'},\delta_\eta]\phi_a = -(\eta'^T\gamma^o\gamma^\mu\eta)P_\mu\phi_a$$

$$- \frac{1}{2}(\eta'^T\gamma^o\sigma^{\mu\nu}\eta)\varepsilon_{\mu\nu\kappa\lambda}p\lambda(i\gamma^\kappa\gamma^5)_{ab}\phi_b \ . \tag{4.4.17}$$

The second term on the right hand side is a "rest frame rotation". This rotation (as well as the factor 2 on the right hand side of (4.4.16)) are removed by considering a second spin zero state of opposite parity. (See Schwinger [SCW 2]). In this way one obtains the standard global supersymmetry result. Whenever, as in (4.4.16) and (4.4.17) one limits the action to a *finite* number of levels, then one obtains *commutation relations* instead of the *anti-commutation relations* satisfied by $\delta_{+\eta}$ and $\delta_{-\eta}$, this is related to another no-go theorem.

Remarks: The use of spinorial operators S_α, $\bar{S}_\beta$ for constructing the algebra is not essential. Instead one may use four-vector shift operators, tensorial shift operators,..., as one chooses. Products of spinorial operators automatically yield higher rank shift operators. It is of interest to note that for vectorial shift operators there exists a variant distinct from that given by the product of two spinorial operators. This variant is characterized by:

$$[\kappa_\mu,\bar{\kappa}_\nu] = (g_{\mu\nu} + P_\mu P_\nu M_{op}^{-2}) \tag{4.4.18}$$

instead of by eq. (4.3.7). Defining $\delta_{+\omega}$ and $\delta_{-\omega}$ in analogy to $\delta_{+\eta}$ and $\delta_{-\eta}$ one can easily construct δ_ω and δ_ν, limited to the levels of spin 0 and spin 1. One finds

$$[\delta_{\omega'},\delta_\omega]\phi_\mu = (g_{\mu\nu} - \frac{P_\mu P_\nu}{m_1^2})(\omega^\nu\omega'^\rho - \omega'^\rho\omega^\rho)\phi_\rho \quad , \tag{4.4.19}$$

$$[\delta_{\omega'},\delta_\omega]\phi_o = 0. \tag{4.4.20}$$

Here we used a four vector index ϕ_μ instead of the pair of symmetric spinor indices ϕ_{ab} to describe the spin one state. Again the term on the right hand side of (4.4.19) is a rest frame rotation of the spin one state. This rotation can again be removed by considering an extra level, (see [SCH 1]).

§5. RELATIVISTIC SU(6)

The construction of an explicitly Poincaré invariant model realizing "relativistic SU(6)" with both mass and spin mixing is now straightforward. One defines an SU(3) multiplet of spinorial operators S_a^i (i = 1,2,3), and generalizes (4.3.7) to read

$$[S_a^j, S_b^k] = \delta^{jk}(-i)\frac{1}{2}(\mathbb{1} + i\gamma_\mu P^\mu M_{op}^{-1})_{ab} \quad , \qquad (4.5.1)$$

and proceeds as before except that the wave function ϕ now contains three pairs of oscillator variables $(\xi_1{}^j, \xi_2{}^j)$, j = 1,2,3. Thus one obtains a manifestly invariant form of the SU(6) structure of Chap. 3, §7. The multiplets associated to this structure are the totally symmetric irreps $[n\dot{0}]$ of SU(6); that is: 1,6,21,56,... . The baryon 56-multiplet is thereby realized; mass-splitting can be incorporated. In Chapter 7 we will demonstrate that minimal electromagnetic coupling generates anomalous magnetic moments which have the desired SU6 symmetry structure.

Since the existence of relativistic SU6 has been regarded as problematic, let us state the geometric meaning of our construction. The symmetry group SU6 is the maximal compact subgroup of the algebraic Lie group generated by the "velocity-Poincaré" generators [eq.(4.3.12)], extended by S_α^i and $\bar{S}_\beta{}^j$.

There have been difficulties in realizing "relativistic SU6" within the standard supersymmetry framework. The incorporation of mass by spontaneous symmetry breaking is a complicated construction. More serious is the fact that anti-symmetric SU(6) multiplets necessarily occur. *Both difficulties disappear in the realization of generalized supersymmetry given above.*

CHAPTER FIVE

CONSTRAINED HAMILTONIAN MECHANICS

In the preceding chapters we constructed the quantum mechanics of a system which can take on all values of spin ("Regge sequence"). This system, however, was free in that it did not have interactions with external fields. Our next task is to introduce interactions with an external electromagnetic field. In Chapter 2 we indicated that coupling to an external electromagnetic field is impossible for the individual levels of spin (it is impossible for the new Dirac equation which describes the spin zero state). It will turn out to be possible to introduce interaction with an external electromagnetic field for the entire system. In order to do this it is best to first go back to a classical Lagrangian theory for our system. The Lagrangian will turn out to be singular and has to be handled by the methods of "Constrained Hamiltonian Mechanics". These methods are elegant and on prolonged acquaintance rather simple and we wish to give a survey of them here.

They have been developed principally by Dirac. We shall try to explain clearly the ideas involved, but omit proofs of the various statements made. For details we refer the reader to the several recent reviews of this subject [DIR 6], [AND 1], [SUD 1], [HAN 1]. We start with the theory and then give some examples.

§1. THEORY

Singular Lagrangians. Let $L\ (q_1,\ldots,q_N,\ \dot q_1,\ldots,\dot q_N) \equiv L(q,\dot q)$ be a Lagrangian for a system with N degrees of freedom. The dot signifies the derivative with respect to an evolution parameter s. The Lagrangian is *singular* if

$$\det\left(\frac{\partial^2 L}{\partial \dot q_i\, \partial \dot q_j}\right) = 0 \quad . \tag{5.1.1}$$

Equation (5.1.1) means that, when the Lagrangian equations of motion are written out in suitable coordinates, the coefficient of a least one $\ddot q$ is zero. Also (5.1.1) means that the equations which give the canonical momenta

$$P_i = \frac{\partial L}{\partial \dot q_i} \quad , \quad i = 1, \ldots n \tag{5.1.2}$$

cannot be solved uniquely for each of the velocities $\dot{q}_1 \ldots \dot{q}_N$ in terms of $q_1 \ldots q_N, p_1 \ldots p_N$. The situation can be visualized in this way. The N dimensional configuration space with coordinates q_r leads automatically to two 2N dimensional spaces: one is the tangent bundle of Lagrangian coordinates and velocities q_r, $\dot{q}_r$, and the other is the cotangent bundle of phase space of coordinates and momenta q_r, p_r. Eqs. (5.1.2) define the Legendre mapping from the former space to the latter. While the domain of this map is always 2N dimensional, for a singular system the range has lower dimension, say $N + M$ for some $M < N$. Over this range, denoted by Σ_o, $q_1, \ldots, q_N, p_1, \ldots, p_N$ are not independently variable but are subject to (N-M) independent constraining relations:

$$\phi_m(q,p) \approx 0 \ , \quad m = 1, \ldots, N - M \ . \tag{5.1.3}$$

The special "weak" equality sign $\approx$ is a reminder that these relations are, of course, not valid over the entire 2N dimensional phase space but only over Σ_o--in fact, they serve to define Σ_o. It is evident that--these relations among q and p arise on eliminating $\dot{q}$ in the N equations (5.1.2). Conversely, if values are given to q and p obeying (5.1.3), ie, if a point of Σ_o is chosen, there will be an (N - M) - fold degree of arbitrariness in the values one can find for $\dot{q}$ so that eqs. (5.1.2) are obeyed. This is because the Legendre map is many-to-one to this extent.

Primary Constraints. The relations (5.1.3) are called "primary constraints" and the region Σ_o the "constraint hypersurface". To exhibit the general solution for q in terms of q, p with the degree of arbitrariness made explicit, one proceeds thus. As in canonical mechanics one sets up the expression

$$H_o = \sum_i p_i\dot{q}_i - L(q,\dot{q}) \tag{5.1.4}$$

It is a property of the Legendre transformation that H is a well-defined function on Σ_o. More precisely, the expression (5.1.4) has the same value at all those points in the $q,\dot{q}$ space that get mapped by (5.1.2) into one point on Σ_o. Thus, (5.1.4) is a function H_o (q,p) on phase space that is unique up to addition of terms that vanish on Σ_o. (For us this will mean that H_o (q,p) is unique up to terms linear in the ϕ_m). The general solution for $\dot{q}_r$ then takes the form

$$\dot{q}_i \approx \{q_i, H_o\} + \{q_i, \phi_m\} v_m, \quad i = 1,\ldots,N. , \tag{5.1.5}$$

that is, the Hamiltonian is: $H = H_o + \phi_m v_m$.

Here the v_m, $m = 1,\ldots,N-M$, are the expected number of arbitrary quantities in the solution, and the Poisson bracket symbol has been introduced. At this point the Poisson bracket is the usual 2N dimensional phase space bracket. The $\approx$ sign is a reminder that in the process of evaluating the Poisson brackets we must treat all the 2N variables q_r, p_r as independent ones; only after the various partial derivatives have been computed must the q's and p's be restricted to Σ_o. It is important, for the validity of (5.1.5), that the forms of the functions ϕ_m be chosen with some care: namely, they must be such that a "first order" displacement δq_r, δp_r taking one out of Σ_o must cause changes of the *same order* in the ϕ's.

The v_m are called *unknown* or *unsolved* velocities. In a particular way of writing H_o and the ϕ_m, they are (N - m) of the $\dot{q}_r$ themselves: for the corresponding values of r, eqs. (5.1.5) then become empty. One may now say that the N equations (5.1.2) of the Legendre map are *completely equivalent in content* to the (N-M) primary constraints (5.1.3) and the N equations (5.1.5) with (N-M) arbitrary v's:

By finding expressions for the generalized forces $\frac{\partial L}{\partial q_r}$ in terms of q, p, v, which is certainly possible since we have eqs. (5.1.5), we can recast the EulerLagrange equations of motion

$$\frac{d}{dt}\frac{\partial L}{\partial \dot{q}_r} - \frac{\partial L}{\partial q_r} = 0 \tag{5.1.6}$$

also in terms of q, p, v_m; they then appear as

$$\dot{p}_i \approx \{p_i, H_o\} + \{p_i, \phi_m\} v_m \quad . \tag{5.1.7}$$

Combining eqs. (5.1.5, 5.1.7) we have an equation of motion for a general phase space function f(q,p,s).

$$\frac{df}{ds} \approx \frac{\partial f}{\partial s} + \{f, H_o\} + \{f, \phi_m\} v_m \quad . \tag{5.1.8}$$

This, together with the set of primary constraints (5.1.3), is the starting point of the analysis of the motion in Hamiltonian language.

We may remark in passing that if L were homogeneous of degree one in $\dot{q}_r$, which happens if we have "chronometric invariance", the function H_o can be taken to be identically zero.

Secondary Constraints. For consistency we must now demand that $\phi_m(q,p)$ remain zero for all s. While the v_m were independent and arbitrary at the level of the Legendre transform it is demands such as these, and similar ones later on, that may to some extent determine or restrict them. Taking $f = \phi_{m'}$ in eq. (5.1.8) we add to the analysis the set of conditions

$$\{\phi_{m'}, H_o\} + \{\phi_{m'}, \phi_m\} v_m \approx 0 \quad . \tag{5.1.9}$$

The possible consequences are of two, and only two, kinds: (i) more constraints in q, p may emerge, thus restricting the motion to some subset of Σ_o, and thereby refining the meaning of the $\approx$ sign; (ii) some of the v_m may be determined in linear inhomogeneous fashion in terms of the others, with functions of q, p as coefficients. Consequences of type (i) are called "secondary constraints" and usually written $\chi_a \approx 0$. Every such χ_a leads in turn to another consistency condition:

$$\{\chi_a, H_o\} + \{\chi_a, \phi_m\} v_m \approx 0 \quad , \tag{5.1.10}$$

which again can imply two kinds of consequences, and so on. Each time we get a reduction in the arbitrariness in the v_m, it effectively results in some function of q, p (in fact some linear combination with q, p dependent coefficients, of the ϕ_m) being added onto H_o in (5.8), and the coefficients of the surviving independent v's will be some linear rearrangement of the primary constraints ϕ_m. At the end of this process, one will have a function H in place of H_o, and some complete set of constraints $\phi_m \approx 0$, $\chi_a \approx 0$. A special set of linear combinations ϕ_α of the ϕ_m, together with arbitrary coefficients v_α, will have been formed, less than (N - M) in number if there has been a reduction in the number of independent v's and the final description of the system's motion will have the form

$$\frac{df}{ds} \approx \frac{\partial f}{\partial s} + \{f,H\} + \{f, \phi_\alpha\} v_\alpha \ ,$$

$$\phi_m \approx 0 \quad , \qquad \chi_a \approx 0 \quad . \tag{5.1.11}$$

We stress that this is a faithful rendering in Hamiltonian language of the original Euler-Lagrange equations of motion and their consequences. The set of all solutions to (5.1.11) thus necessarily coincides with the set of all solutions to (5.1.6). To fix a solution to (5.1.11) involves giving a set of initial values for all the $\underline{q}$'s and p's consistent with the constraints, and specifying the v_α for all s in some way.

First Class and Second Class Constraints. Let the constraint hyper surface determined by the ϕ_m and χ_a together be denoted by Σ. Completeness of the constraints, and freedom in the choice of the v_α, imply the following: the infinitesimal changes in q, p produced by *each* of the PB's in (5.1.11), in an interval δs, has the property of mapping Σ_o onto itself. Thus this is a property characterizing H and the ϕ_α. A phase space function f is said to be *first class* if the canonical transformation generated by it preserves Σ_o, otherwise it is *second class*. So H and ϕ_α are first class: in fact the ϕ_α are a maximal set of independent first class linear combinations of ϕ_m, each present in the final equation of motion with its own arbitrary coefficient. The functions ϕ_m, χ_α are handled in two ways: on the one hand through their vanishing they determine the region Σ_o to which the motion is restricted; on the other hand each of them (or any linear combination) generates a canonical transformation whose effect on Σ_o tells us if it is first or second class. This is indicated in figure 5.1, where we have labelled all first class constraints by η and all second class constraints by θ. The action associated with an η is a canonical transformation within the surface whereas θ generates a canonical transformation which leads out of the surface. Figure 5.1 is qualitative as it is impossible to draw it in sufficient dimensions.

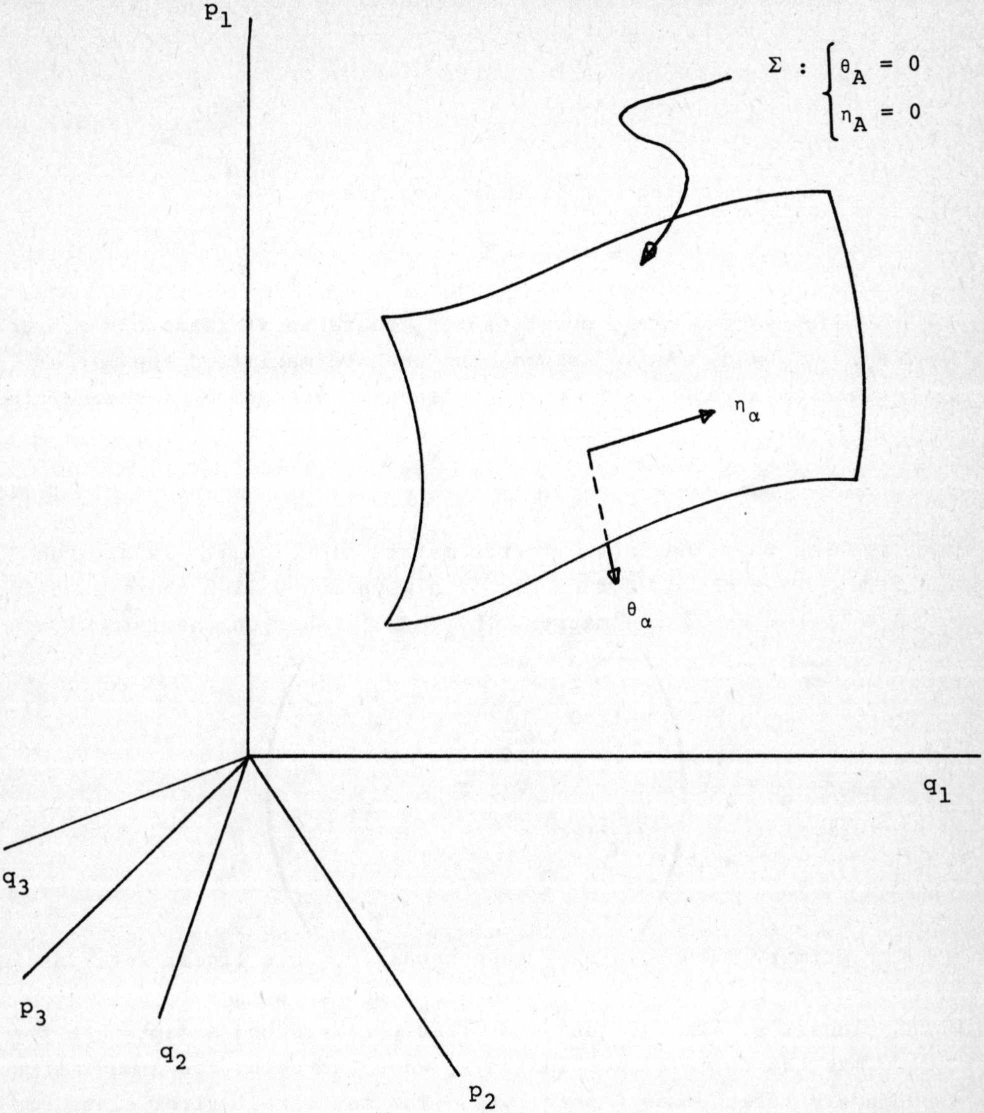
p_1
$\Sigma : \begin{cases} \theta_A = 0 \\ \eta_A = 0 \end{cases}$
η_α
θ_α
q_1
q_3
p_3
q_2
p_2

Figure 5.1

The first class constraints $\leftrightarrow, \eta_\alpha$, must satisfy

$$\{\eta_\alpha, \eta_\beta\} \approx 0 \text{ , for all } \alpha,\beta, \tag{5.1.12}$$

$$\{\eta_\alpha, \theta_A\} \approx 0 \quad \text{for all } \alpha, A, \tag{5.1.13}$$

whereas by the preceding construction one has also

$$\{\eta_\alpha, H\} \approx 0, \quad \{\theta_A, H\} \approx 0. \tag{5.1.14}$$

For each second class constraint θ_A there is at least one θ_B such that $\{\theta_A, \theta_B\} \neq 0$. As we assume that we have maximized the set of first class constraints, there is no linear combination of second class constraints which is first class. Therefore

$$\det \quad \|\{\theta_A, \theta_B\}\| \not\approx 0 \tag{5.1.15}$$

which implies that the antisymmetric matrix $\|\{\theta_A, \theta_B\}\|$ is of even dimension. Thus the number of θ's is always even, also there is a coordinate system where the matrix $\|\{\theta_A, \theta_B\}\|$ takes on the form

$$\|\{\theta_A, \theta_B\}\| = \begin{pmatrix} 0 & 1 & & & \\ -1 & 0 & & & \cdots \\ & & 0 & 1 & \cdots \\ & & -1 & 0 & \\ \cdot & & & & \\ \cdot & & & & \\ \cdot & & & & \end{pmatrix} \tag{5.1.16}$$

The primary constraints ϕ_m thus separate, as a linear set, into a maximal number of primary first class ϕ_α, and a balance of primary second class ϕ's. The χ_a can be similarly rearranged except that for them we may add terms linear in the ϕ_m when searching for a maximal set of secondary first class constraints. The set of all first class constraints, primary ϕ_α plus secondary ones, are the η's above. All remaining second class constraints make up the θ's.

Second class Constraints: induced metric. There is an important physical difference between first class constraints and second class constraints. To begin with second class constraints, these occur in pairs as is obvious from the normal form (5.1.16) of $\{\theta_A, \theta_B\}$. Each pair corresponds, locally to a pair of conjugate variables, p and q in suitable coordinates which should be eliminated as canonical variables. In suitable local coordinates the second class constraints are typically of the form $p_1 = 0$, $q_1 - 10 = 0$. Hence on Σ_o p_1 and q_1 should be eliminated and this leads to a modification of the Poisson bracket as now on Σ, p, and q, will bracket zero. This new bracket is what is called the Dirac bracket. What happens is quite analogous to what happens when one restricts oneself to a surface in n dimensional Euclidean space. The n-dimensional Euclidean metric induces a metric in the surface. Again locally this is quite simple; let the surface be n-1 dimensional then in a suitable coordinate system this takes the form that the first coordinate is fixed and the metric is the n-1 dimensional metric in the remaining coordinates. Here we are instead in a 2n dimensional phase space with a 2n dimensional antisymmetric metric $\{q_N, p_M\} = \delta_{NM}$. The surface Σ_o will for simplicity be chosen 2n-2 dimensional. Again the 2n dimensional metric induces in Σ_o a metric. This metric is that of the Dirac Bracket. Locally this Dirac bracket is simple, in suitable coordinates it means that p and q are fixed and eliminated from the bracket. We shall give a closed expression for this Dirac bracket after finishing our comparisons of first and second class constraints.

First Class Constraints: gauge transformations. The *first class constraints* correspond to *gauge transformations*. To be precise: Let us assume that there are only primary and secondary constraints, then to every $\eta\alpha$ there corresponds an invariance transformation of the action $W = \int ds\, L$:

$$\delta q_1 = \varepsilon(s)\ \delta^{(1)} q_i + \dot{\varepsilon}(s)\ \delta^{(2)} q_i \quad , \tag{5.1.17}$$

where $\dot{\varepsilon}(s) = \frac{d}{ds}\varepsilon(s)$, with $\varepsilon(s)$ an arbitrary function, and where $\delta^{(1)} q_i$ and $\delta^{(2)} q_i$ are specified functions of the q's, $\dot{q}$'s and of s. If $\delta^{(2)}\ q_i = 0$ for all i, then the first class constraint which generates (5.1.17) is a linear combination of (first class) primary constraints only. When $\delta^{(2)}\ q_i \neq 0$ for some i the generator of (5.1.17) contains at least one (first class) secondary constraint.

Dirac Brackets. Now let us give a global expression for the Dirac bracket. As stated before this bracket is the antisymmetric metric induced on Σ_o by the antisymmetric metric in the original 2n dimensional phase space ($\{p_N, q_M\} = -\delta_{NM}$, etc.). In suitable local coordinates this induction is quite simple, it corresponds to the elimination of q_1 and p_1 (and possibly several other such pairs) as canonical variables. Consider again the antisymmetric nonsingular matrix composed from the second class constraints considered in

(5.1.15): $\|\{\theta_A, \theta_B\}\|$. Let the inverse of this matrix be C_{AB}:

$$C_{AB} \,\|\{\theta_B, \theta_C\}\| = \delta_{AC} \ . \tag{5.1.18}$$

Then the global expression for the induced bracket (Dirac bracket) is:

$$\{f,g\}^* = \{f,g\} - \{f,\theta_A\}\, C_{AB}\, \{\theta_B, g\} \tag{5.1.19}$$

That (5.1.19) is correct can be seen from the fact that it does not depend on canonical coordinates nor on the fact that in suitable coordinates $\|\{\theta_A, \theta_B\}\|$ takes on the form (5.1.16). The suitable coordinates of the last sentence can only be chosen pointwise on Σ_o, unless Σ_o is "flat".

The basic properties of the DB are: (i) it has the same linearity, antisymmetry and derivation properties as the PB; (ii) it obeys the Jacobi identity, (iii) $\{\theta_A, g\}^*$ vanishes identically for any g; (iv) for a first class f, $\{f,g\}^*$ and $\{f,g\}$ are weakly equal. Because of property (iv) the PB's in (5.1.11) may be replaced by DB's. Once this is done, one can use the vanishing of the θ_A as identities and eliminate the corresponding number of q's and p's. Only the first class constraints need to be stated explicitly as accompanying the general equation of motion in DB form.

Gauge Constraints, Invariant Relations. Two important concepts remain to be explained in connection with the development of the dynamical equations: they relate to gauge constraints and invariant relations. The theory based on the Lagrangian, in its final form, involves the arbitrary quantities v_α in eq. (5.1.11). Gauge constraints χ_α' (qps) ≈ 0 are a set of constraints imposed to fix the v_α: we need as many χ_α' as there are primary first class ϕ_α, and the two together must be a second class system (with respect to the DB (5.1.13) but actually equally well respect to the PB). This is assured if

$$\det |\{\chi_\alpha', \phi_\beta\}^*| \not\approx 0 \ . \tag{5.1.20}$$

It is important to recognize that the χ_α' do not arise from the action principle based on L but are "chosen from the outside". If $(A_{\alpha\beta})$ is the inverse to the matrix appearing in (5.1.20),

$$A_{\alpha\beta}\, \{\chi_\beta', \phi_\gamma\}^* \approx \delta_{\alpha\gamma} \ , \tag{5.1.21}$$

the conservation of the χ_α' determine the v_α:

$$\frac{d\chi'_\alpha}{ds} = \frac{\partial\chi'_\alpha}{\partial s} + \{\chi_\alpha', H\}^* + \{\chi_\alpha', \phi_\beta\}^* v_\beta \approx 0 \quad \Rightarrow$$

$$v_\alpha = - A_{\alpha\beta}\left(\frac{\partial\chi_\beta'}{\partial s} + \{\chi_\beta', H\}^*\right) \tag{5.1.22}$$

We may point out that since H, ϕ_α are all first class, the DB's in eqs. (5.1.20, 5.1.21, 5.1.22) can be replaced by PB's. If we wish we may at this stage pass from the DB's $\{\ ,\ \}^*$ to a final set of DB's $\{\ ,\ \}^{**}$ arising on elimination of the ϕ_α and χ_α' :

$$\{f,g\}^{**} = \{f,g\}^* - A_{\alpha\beta}(\{f,\phi_\alpha\}^*\{\chi'_\beta, g\}^* - \{f,\chi_\beta'\}^*\{\phi_\alpha, g\}^*)$$

$$- \{f,\phi_\alpha\}^* A_{\alpha\beta}\{\chi'_\beta, \chi'_{\beta'}\} A_{\alpha'\beta'}\{\phi_{\alpha'}, g\}^* \quad . \tag{5.1.23}$$

The way to handle the resulting equation of motion, which has unfamiliar terms steming from the explicit s-dependence in χ_α', is explained in [MUK 2]

Dirac has suggested that under certain circumstances it may be appropriate to generalize the equation of motion (5.1.11) by adding terms proportional to the secondary first class constraints, each with a free coefficient. If this is done, the resulting system definitely goes beyond the Lagrangian theory: every solution of the Lagrangian system is a solution of the enlarged one but not conversely. For the enlarged theory one must then choose as many gauge constraints as the total number of first class constraints the Lagrangian led to, if all the arbitrary v's are to be fixed.

Invariant relations are relations Ψ_λ (q,p) $\approx$ 0, say, which one may impose at any stage to limit the class of solutions of the equations of motion one wants to consider. Their important property is that they are self-perpetuating, ie, if obeyed at an initial instant they are obeyed throughout the motion. At whatever stage they may be adopted, they do not and are not intended to fix any velocities v that are arbitrary at that stage. Thus invariant relations effectively limit the possible choices of initial conditions only.

We conclude this brief summary of constrained Hamiltonian dynamics with three examples.

§ 2. EXAMPLES

We give three examples, the first involves second class constraints only, the second concerns a comparison of these two classes of constraints, and the third shows clearly the relation between first class constraints and gauge transformations.

Example 1. The first example is one with second class constraints only, in fact we meet this example in a slightly more general form in Chapter 7. Consider the Lagrangian

$$L = \dot{q}_1 q_2 - \dot{q}_2 q_1 + q_1^2 + q_2^2 \quad , \qquad (5.2.1)$$

one finds

$$p_1 = \frac{\partial L}{\partial \dot{q}_1} = q_2 \ , \quad p_2 = \frac{\partial L}{\partial \dot{q}_2} = - q_1 \quad ,$$

i.e. with the Poisson brackets $\{q_i, p_j\} = \delta_{ij}$ we have two primary constraints:

$$\phi_1 = p_1 - q_2 \approx 0$$
$$\phi_2 = p_2 + q_1 \approx 0 \qquad (5.2.2)$$

with

$$\{\phi_1, \phi_2\} = -2 \quad .$$

The Hamiltonian (5.1.5) is

$$H = -(q_1^2 + q_2^2) + v_1(p_1 - q_2) + v_2(p_2 + q_1) \ .$$

One now demands $\dot{\phi}_1 \approx 0, \quad \dot{\phi}_2 \approx 0$, i.e.

$$0 \approx \{H, \phi_1\} = - 2q_1 + 2v_2$$

$$0 \approx \{H, \phi_2\} = 2q_2 - 2v_1 \quad .$$

The equations imply $v_1 = q_2, v_2 = q_1$ and no further (secondary) constraints. Thus the only constraints are (5.2.2) and these are second class. We have

$$\left(\{\phi_i, \phi_j\}\right)^{-1} = \frac{1}{\sqrt{2}} \begin{pmatrix} 0 & 1 \\ -1 & 0 \end{pmatrix} ,$$

corresponding to (5.1.18). The Dirac brackets are: $\{q_1, q_2\}^* = 1$ and the constraints become identities: $p_1 = q_2, \ p_2 = -q_1$, and the Hamiltonian becomes $H = -(q_1^2 + q_2^2)$, with $\{q_1, q_2\}^* = 1$. Now, this is nothing but the Hamiltonian for the 1-dimensional harmonic oscillator with its usual Poisson bracket, except that what one usually writes as p_1 is denoted by q_2 here. Thus the Lagrangian (5.2.1) is nothing but a disguised Lagrangian which really describes a 1-dimensional harmonic oscillator.

Example 2. Our second example refers to classical field theory, where the p's and q's carry a continuous label. We discuss briefly the Lagrangian for classical vector fields and compare the mass zero equations (Maxwell) with the Proca equations for a massive field. The Maxwell field has been discussed in detail by Dirac [DIR 6] and the comparison of Maxwell and Proca fields is also carried out in [SUD 2] hence we shall only give a brief survey here. We give both Lagrangians at the same time, putting the extra mass term of the Proca field in brackets:

$$L = \int d\vec{x}\ [-\tfrac{1}{4}F_{\mu\nu}F^{\mu\nu} + (+\tfrac{1}{2}m^2A_\mu A^\mu)] \qquad (5.2.3)$$

where $F_{\mu\nu} = \partial_\mu A_\nu - \partial_\nu A_\mu$, one finds for $\Pi^\mu(\vec{x})$, the conjugate of $A_\mu(\vec{x})$,

$\Pi^\mu(\vec{x}) = \dfrac{\partial L}{\partial \dot{A}_\mu(\vec{x})} = F^{\mu o}(\vec{x})$, i.e. for both $m = 0$ and $m \neq 0$ $\Pi^0(\vec{x}) = 0$.

Writing the Poisson brackets

$$\{A_\mu(\vec{x}), \Pi^\nu(\vec{y})\} = g_\mu{}^\nu\delta(\vec{x} - \vec{y}) , \qquad (5.2.4)$$

one has the primary constraint

$\Pi^0(\vec{x}) \approx 0$; the Hamiltonian is

$$H = \int d\vec{x}\ [\Pi^\mu A_{\mu,o} + \tfrac{1}{4}F_{\mu\nu}F^{\mu\nu} + (-m^2A_\mu A^\mu) + v_1\Pi^0(\vec{x})] =$$

$$= \int d\vec{x}\ [\tfrac{1}{4}F_{rs}F^{rs} + \tfrac{1}{2}\Pi^r\Pi_r - A_o\Pi^r{}_{,r} + (-\tfrac{1}{2}m^2A_\mu A^\mu) + v_1\Pi^0],$$ with it

one finds the secondary constraint: $\{H, \Pi^0\} = \Pi^r{}_{,r} + (-m^2A_o) \approx 0$. At this point a qualitative difference occurs between the cases $m = 0$ (Maxwell) and $m \neq 0$ (Proca).

Proca. First, take $m \neq 0$ the constraints are:

$$m \neq 0 \quad \begin{array}{l} \Pi^o = \phi_1(\vec{x}) \approx 0 \\ A_o - \dfrac{1}{m^2}\Pi^r{}_{,r} = \phi_2(\vec{x}) \approx 0 \end{array} \qquad (5.2.5)$$

, and with (5.2.4)

$\{\phi_1(\vec{x}),\phi_2(\vec{y})\} = \delta(\vec{x} - \vec{y})$, i.e. the

constraints (5.2.4) are second class and they lead to the algebraic elimination of one of the four degrees of freedom at every $\vec{x}$. The constraints become equations expressing the 0 components algebraically in the others:

$$\Pi^o = 0$$

$$A^o = \frac{1}{m^2}\,\Pi^r{}_{,r}$$

and the Dirac Brackets take care of this reduction. By the same token, the consistency condition $\dot{\phi}_2 \approx 0$ actually results in v_1 being determined, so no free coefficients remain in H. Note that one needs two constraints at every point $\vec{x}$ to eliminate one of the initially four Lagrangian degrees of freedom!

Maxwell. Next, let us consider the m = 0 case, the constraints are then:

$$m=0 \qquad \begin{aligned} \Pi^o &= \phi_1 \approx 0 \\ \Pi^r{}_{,r} &= \phi_2 \approx 0. \end{aligned} \qquad (5.2.6)$$

Unlike (5.2.5) these constraints are with (5.2.4) first class:

$$\{\phi_1, \phi_2\} = 0.$$

Hence, instead of algebraic elimination of one of the degrees of freedom we now have two generators (Π^o and $\Pi^r{}_{,r}$) of gauge transformations and in the Hamiltonian H, v remains free. Each of these two generators eliminates a degree of freedom at every point $\vec{x}$, reducing the number of degrees of freedom at every $\vec{x}$ from four to the two characteristic of photons. Thus there is actually a *discontinuity* in what happens for $m \neq 0$ and what happens for m = 0. This discontinuity is more serious for the case of symmetric spin two tensor fields. For these classical spin two fields one finds a discontinuity in the results for classical theories of gravity [VAN 6] [FOR 1]. This is the discontinuity which was discovered before in the context of quantum field theory [VAN 7]. Briefly: although the Lagrangian appears continuous for $m \to 0$, as in (5.2.3), the Lagrangian is actually a singular Lagrangian and this singularity is treated *differently* for m = 0 (gauge invariance) from $m \neq 0$, $m \to 0$ (algebraic elimination of variables), and this leads to a discontinuity in the retarded propagator.

Example 3 Point Particle. This example is treated in detail in [HAN 1], nevertheless we shall go over it briefly as it is of great importance in relation to the next two chapters. It concerns the equation for the world line of a single point particle, as drawn in figure 5.2.

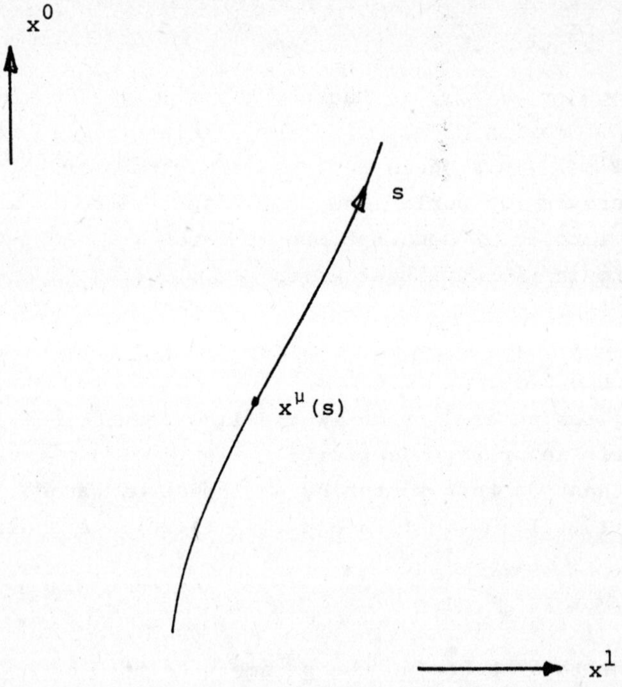

Figure 5.2

The parameter s is rather arbitrary. Let $x^\mu(s) = \frac{d}{ds} x^\mu(s)$. Then one may write the Lagrangian

$$L = m_o \sqrt{-\dot{x}^2(s)} \ . \tag{5.2.7}$$

with action $W = \int ds\, L\,(s)$. This Lagrangian is invariant for *chronometric* transformations

$$s \rightarrow f(s) \quad s' . \tag{5.2.8}$$

These are gauge transformations, as an arbitrary function is involved. Therefore the Lagrangian (5.2.7) is singular and there is a constraint which generates the gauge transformations (5.2.8). One has:

$$P_\mu = \frac{\partial L}{\partial \dot{x}^\mu} = - \frac{m_o \dot{x}_\mu}{\sqrt{-\dot{x}2}} \ .$$

This implies the constraint

$$P^\mu P_\mu + m_o^{\ 2} \approx 0 \ . \tag{5.2.9}$$

It is this constraint which generates the transformations (5.2.7). Clearly also

$$\frac{\partial^2 L}{\partial \dot{x}^\mu \partial \dot{x}^\nu} = \frac{-1}{\sqrt{-\dot{x}^2}} [g_{\mu\nu} - \frac{\dot{x}_\mu \dot{x}_\nu}{\dot{x}^2}] \quad , \tag{5.2.10}$$

which is a singular matrix, in Lagrangian language this implies that the equations of motion do not give the acceleration in the direction of the world line. This again reflects the chronometric invariance: all that matters is the world line, how "fast" x_μ moves with s is irrelevant as long as $\dot{x}_\mu$ does not become zero.

The Hamiltonian for the system is

$$H = 0 + v(P^2 + m_o^2) \tag{5.2.11}$$

with this Hamiltonian the motion (world line) can be solved leaving the chronometric invariance explicit.

One often handles this system by a "choice of gauge", as discussed in general in (5.§1) just before equation (5.20). A choice of gauge is

$$0 \approx \chi = x^o - s \quad , \tag{5.2.12}$$

this then leads to x^o becoming the parameter s and to the Dirac bracket

$$\{P^\mu, x^\nu\}^* = -g^{\mu\nu} + g^{\mu o} \frac{P^\nu}{P^o} \quad . \tag{5.2.13}$$

CHAPTER SIX

VECTOR LAGRANGIAN MODEL

In the preceding chapters we have discussed the quantum mechanics of free relativistic systems which have an internal structure of the kind introduced in the new Dirac equation.

The introduction of interaction, say with an external electromagnetic field, is not obvious for the quantum mechanical model of Chapters 3 and 4 . This circumstance makes us look at the problem anew, this time with a classical Lagrangian starting point [MUK 3]. The basic model of Chapters 3 and 4 exhibits a spin spectrum with each integer and half integer value of spin occurring just once. The classical Lagrangian which describes this model is discussed in Chapter 7, where interaction is also introduced and discussed in some detail. For purposes of orientation and also because of its intrinsic interest we develop in the present chapter a classical model which upon quantization realizes the sequence of integer spin values. This is the vectorial model.

§1. CHOICE OF VARIABLES AND LAGRANGIAN

We want to describe a classical relativistic indecomposable system with the following property: if we were to imagine a quantized version of the system, in the rest frame each integer angular momentum value ℓ = 0, 1, 2, must appear just once. Among the classical configuration variables, we must in any case include a Lorentz four-vector x^μ to describe space time location. The canonical formalism then automatically ensures the existence of a canonically conjugate variable P_μ which will be the total four momentum provided all other variables are translation invariant.

Variables. What are the additional variables we must introduce? It is clear that a genuine two-particle system, with x^μ being the center in some sense, would be inappropriate: in the rest frame radial motions are possible and in the quantum theory each integer ℓ would occur infinitely many times. To get the multiplicity-free spectrum ℓ = 0, 1, 2, ..., we must choose an internal variable which reduces, in the rest frame, to a unit spatial vector with arbitrary orientation. One way of implementing this in a covariant way is to introduce a unit space like translation-invariant four-vector a^μ for which we ensure the constraint $P.a \approx 0$.

Our object thus has a world-line $x^\mu(s)$, where s is an arbitrary parameter labelling the events along the world-line, and in addition at each s a vector $a^\mu(s)$ obeying $a^\mu(s)\, a_\mu(s) = 1$. The configuration space is seven dimensional, with generalized coordinates, x^μ, a^μ. Under an element (Λ, d) of the Poincaré group, they transform as

$$x^\mu(s) \to x'^\mu(s) = \Lambda^\mu{}_\nu x^\nu(s) + d^\mu \quad ,$$
$$a^\mu(s) \to a'^\mu(s) = \Lambda^\mu{}_\nu a^\nu(s) \tag{6.1.1}$$

Differentiation with respect to s will be indicated by a dot, so

$$a^2 = 1 \text{ implies } a.\dot{a} = 0 \tag{6.1.2}$$

The Lagrangian L will be permitted to depend on x, a and their first derivatives $\dot{x}$, $\dot{a}$ alone.

Conditions on L. We impose now three requirements: (i) L must be invariant under the action (6.1.1) of the Poincaré group; (ii) as a result of the definition of P_μ one must produce the constraint $P.a \approx 0$; (iii) the action $\int L ds$ must be unaltered by the replacement of s by a monotonic function s'(s) of itself. Condition (i) tells us first that due to translation invariance L is a function of $\dot{x}$, a, $\dot{a}$; and second that due to homogeneous Lorentz invariance it is a function of the four independent scalars $\dot{x}^2$, $\dot{x}.a$, $\dot{x}.\dot{a}$ and $\dot{a}^2$. Given such an L, the momentum conjugate to x^μ is defined by

$$P_\mu = \frac{\partial L}{\partial \dot{x}^\mu} = \left(2\dot{x}_\mu \frac{\partial}{\partial(\dot{x}^2)} + a_\mu \frac{\partial}{\partial(\dot{x}.a)} + \dot{a}_\mu \frac{\partial}{\partial(\dot{x}.\dot{a})}\right) L \tag{6.1.3}$$

(The problem of defining a suitable canonical conjugate to a^μ will be taken up in §2). Requirement (ii) leads to a first order partial differential equation for L:

$$P.a \approx 0 \text{ gives } \left(\frac{\partial}{\partial(\dot{x}^2)} + \frac{\partial}{\partial(\dot{x}.a)^2}\right) L = 0 \tag{6.1.4}$$

Then both (i) and (ii) are satisfied by permitting L to be an arbitrary function of the three independent scalars

$$u_1 = (\dot{x}.a)^2 - \dot{x}^2 \ , \quad u_2 = \dot{x}.\dot{a} \ , \quad u_3 = \dot{a}^2 \ . \tag{6.1.5}$$

One must appreciate the fact that we have been able to secure the constraint $P.a \approx 0$ so easily, merely by restricting the variables that may enter as arguments of L, because this constraint is linear in the conjugate momenta.

Let us now turn to the third requirement on L. We shall call this *chronometric invariance* in place of the more usual but rather ungainly term *reparametrization invariance*. This condition implies that L is homogeneous of degree one in the velocities $\dot{x}$, $\dot{a}$, or of degree $\frac{1}{2}$ in the u's.

Lagrangian. Thus the most general Lagrangian is of the form

$$L = \sqrt{u_1}\, f\,(\zeta, \eta) \,, \tag{6.1.6}$$

where

$$\zeta = \frac{u_1}{u_2} \,, \qquad \eta = \frac{u_1}{u_3} \,, \tag{6.1.7}$$

and where f is an arbitrary real function of two variables.

Chronometric Invariance. This invariance expresses the following physical idea: even though for the sake of manifest relativistic invariance we introduce all four components x^μ as dynamical variables, ultimately the freedom in choice of parameter s allows us, if we wish, to turn x^o into a parameter and treat x^j alone as true dynamical variables. Thus this property is in accord with the counting of true physical degrees of freedom. On the other hand it leads to the appearance of the cumbersome square root in eq. (6.1.6), apart from any other radicals that may be present in a specific f. Such a situation arises even for a structureless point particle whose Lagrangian would be $\sqrt{\dot{x}^2}$ as discussed in Chapter 4, example 3. For this case it has been known for a long time that for some purposes one may use instead the more manageable expression $\dot{x}^2$ as Lagrangian; even though the counting of degrees of freedom goes awry, it is more convenient for path integral quantization. A similar proposal has recently been made for the relativistic string [EGU 1]. Both for the point particle and the string, the chronometric invariant action has a geometrical meaning: length of arc and area of sheet. Our Lagrangian (6.1.6) or rather the associated action, is not intended to have an intrinsic geometric meaning, and we shall insist on chronometric invariance.

The Lagrangian (6.1.6) thereby leads to (at least) two constraints, one being $P.a \approx 0$ and the other resulting from chronometric invariance. The methods of Dirac's generalized Hamiltonian dynamics recounted in Chapter 5 must therefore be used to analyze the system. All this is very much analogous to the model of Regge and Hanson discussed in Appendix A.

§2. PHASE SPACE STRUCTURE AND CONSERVATION LAWS

Among the four variables a^μ just three are algebraically independent. In a phase space formulation of the system defined by the Lagrangian (6.1.6) there must therefore appear just three algebraically independent canonical conjugates to a^μ.

We need a covariant version of the canonical formalism respecting the condition $a^2 = 1$. Suppose we had a Lagrangian $L_1(a, \dot{a})$ in which all four components of a and $\dot{a}$ appear explicitly. (For the moment we suppress other variables that may be present). We might eliminate a_o and $\dot{a}_o$ right away using the equations

$$a_o = \sqrt{a_j a_j - 1} \quad , \qquad \dot{a}_o = \frac{a_j \dot{a}_j}{a_o} \quad . \tag{6.2.1}$$

Substituting these in L_1 we get a Lagrangian $L_2(a_j, \dot{a}_j)$ involving the truly independent coordinates and velocities only. The normal rules of canonical mechanics then lead us to define a conjugate variable Π_j and Poisson brackets (PB) as follows:

$$\Pi_j = \frac{\partial L_2}{\partial \dot{a}_j} \tag{6.2.2a}$$

$$\{a_j, a_k\} = \{\Pi_j, \Pi_k\} = 0 \ , \quad \{a_j, \Pi_k\} = \delta_{jk} \quad . \tag{6.2.2b}$$

Let us rewrite Π_j in terms of L_1 and compute the PB's involving a_o:

$$\Pi_j = \frac{\partial L_1}{\partial \dot{a}_j} + \frac{a_j \partial L_1}{a_o \partial \dot{a}_o} \ , \tag{6.2.3a}$$

$$\{a_o, a_j\} = 0, \quad \{a_o, \Pi_j\} = \frac{a_j}{a_o} \tag{6.2.3b}$$

Fully Covariant Treatment. We are looking for a canonical formalism which does not insist on explicit elimination of a_o, $\dot{a}_o$ and so works directly with L_1. Now the natural infinitesimal motions that a_μ is subject to are the infinitesimal transformations of SO(3,1), so we search for functions $S_{\mu\nu}$ of a_j and Π_j such that

$$\{S_{\mu\nu} \ , \ a_\lambda\} = g_{\mu\lambda}\, a_\nu - g_{\nu\lambda}\, a_\mu \quad . \tag{6.2.4}$$

With the help of eqs. (6.2.2b, 6.2.3b) we find:

$$S_{jk} = a_j \Pi_k - a_k \Pi_j \quad , \quad S_{oj} = a_o \Pi_j \tag{6.2.5}$$

But this can be expressed directly in terms of L_1 in a neat form:

$$S_{\mu\nu} = a_\mu \frac{\partial L_1}{\partial \dot{a}^\nu} - a_\nu \frac{\partial L_1}{\partial \dot{a}^\mu} \tag{6.2.6}$$

and from eqs. (6.2.2, 6.2.3b, 6.2.5) the PB's among the $S_{\mu\nu}$ turn out to have the anticipated values:

$$\{S_{\mu\nu}, S_{\rho\sigma}\} = g_{\mu\rho} S_{\nu\sigma} - g_{\nu\rho} S_{\mu\sigma} + g_{\mu\sigma} S_{\rho\nu} - g_{\nu\sigma} S_{\rho\mu} . \tag{6.2.7}$$

It may now be seen that in a fully covariant treatment we can *define* $S_{\mu\nu}$ to be the entity canonically conjugate to a_μ: it is given in terms of L_1 by eq. (6.2.6) with all components of a, $\dot{a}$ treated *as though* they were independent, and the basic PB's are taken to be eqs. (6.2.4)(6.2.7), along with the vanishing of $\{a_\mu, a_\nu\}$. That there are only three algebraically independent variables among the $S_{\mu\nu}$ is guaranteed by the identities

$$a_\lambda S_{\mu\nu} + a_\mu S_{\nu\lambda} + a_\nu S_{\lambda\mu} = 0 \quad . \tag{6.2.8}$$

The Hamiltonian is a function $H(a_\lambda, S_{\mu\nu})$ on the six dimensional generalized phase space and is given by

$$H(a,S) = \Pi_j \dot{a}_j - L_2 = \dot{a}_\mu \frac{\partial L_1}{\partial \dot{a}_\mu} - L_1 \quad . \tag{6.2.9}$$

For the Hamilton-Jacobi theory one makes the substitution

$$S_{\mu\nu} = a_\mu \frac{\partial S(a)}{\partial a^\nu} - a_\nu \frac{\partial S(a)}{\partial a^\mu} \quad . \tag{6.2.10}$$

Vector Variable Conjugate to a_μ. While the above is a natural generalization of the canonical formalism to accommodate the condition $a^2 = 1$, *it is possible to replace the tensor* $S_{\mu\nu}$ *by a simpler object, a vector* b_μ, *as the canonical conjugate to* a_μ. If we view a_μ, $S_{\mu\nu}$ as basic, we may define b_μ by

$$b_\mu = S_{\mu\nu} a^\nu \quad , \tag{6.2.11}$$

and we discover the following algebraic relations and PB's:

$$b_\mu = a_\mu a_\nu \frac{\partial L_1}{\partial \dot{a}_\nu} - \frac{\partial L_1}{\partial \dot{a}^\mu} \ , \qquad \text{(a)}$$

$$a.b = 0 \qquad \text{(b)} \qquad (6.2.12)$$

$$\{a_\mu, b_\nu\} = a_\mu a_\nu - g_{\mu\nu}, \ \{b_\mu, b_\nu\} = a_\mu b_\nu - a_\nu b_\mu . \qquad \text{(c)}$$

However, since with the help of eq. (6.2.8) we are able to invert eq. (6.2.11) and express $S_{\mu\nu}$ in terms of a_μ and b_μ,

$$S_{\mu\nu} = b_\mu a_\nu - b_\nu a_\mu \ , \qquad (6.2.13)$$

we can reverse the roles of $S_{\mu\nu}$ and b_μ and regard the latter as the basic, the former as the derived, quantity. In this view, b_μ is defined by eq. (6.2.12a), has just three algebraically independent components because of eq. (6.2.12b), and the fundamental PB's are in eq. (6.2.12c) (along with the vanishing of $\{a_\mu, a_\nu\}$).

The Hamiltonian and the Hamilton-Jacobi recipe for b_μ are:

$$H(a,b) = - b_\mu \dot{a}^\mu - L_1,$$
$$b_\mu = - \frac{\partial S(a)}{\partial a^\mu} + a_\mu a_\nu \frac{\partial S(a)}{\partial a_\nu} \ . \qquad (6.2.14)$$

It is amusing to notice that the PB's among $S_{\mu\nu}$ and b_μ, namely eq. (6.2.7) together with

$$\{S_{\mu\nu}, b_\lambda\} = g_{\mu\lambda} b_\nu - g_{\nu\lambda} b_\mu \ ,$$
$$\{b_\mu, b_\nu\} = - S_{\mu\nu} \ , \qquad (6.2.15)$$

realize the Lie algebra of the de Sitter Group SO(3,2). This is an example of a well-known construction (BOH-1). A useful result is:

$$b^2 = \frac{1}{2} S_{\mu\nu} S^{\mu\nu} \ . \qquad (6.2.16)$$

Lagrangian, General Case. We now return to the Lagrangian (6.1.6) and its phase space analysis. Since the configuration space is seven dimensional, the phase space Γ is of dimension fourteen and is spanned by the four-vector variables x^μ, a^μ, P^μ, b^μ. The two relevant conditions $a^2 = 1$, $a \cdot b = 0$ are to be counted as kinematic conditions, not constraints in the Dirac sense. For convenience we list again the definition of P_μ and b_μ for L of the form (6.1.6), and give the complete set of non-vanishing PB's over Γ:

$$P_\mu = \frac{1}{\sqrt{u_1}}\left(f+2\zeta\frac{\partial f}{\partial\zeta} + 2\eta\frac{\partial f}{\partial\eta}\right)(\dot{x}.a\, a_\mu - \dot{x}_\mu) - \frac{1}{\sqrt{u_1}}\zeta^2\frac{\partial f}{\partial\zeta}\dot{a}_\mu \quad ,$$

$$b_\mu = -\frac{1}{\sqrt{u_1}}\zeta^2\frac{\partial f}{\partial\zeta}(\dot{x}.a\, a_\mu - \dot{x}_\mu) + \frac{2}{\sqrt{u_1}}\eta^2\frac{\partial f}{\partial\eta}\dot{a}_\mu \quad ; \qquad \text{(a)}$$

$$\{x_\mu, P_\nu\} = g_{\mu\nu},\ \{a_\mu, b_\nu\} = a_\mu a_\nu - g_{\mu\nu},\ \{b_\mu, b_\nu\} = a_\mu b_\nu - a_\nu b_\mu . \qquad \text{(b)} \tag{6.2.16}$$

As we are dealing with a constrained system it is preferable to view the full phase space Γ as intrinsically defined by the configuration space of x and a: Γ is the cotangent bundle over the configuration space. Equations (6.2.16a) are interpreted as giving the Legendre mapping from the space of Lagrangian variables $x, a, \dot{x}, \dot{a}$ (the tangent bundle) into Γ. The singularity of L causes the range of this mapping to be a proper subset of Γ. This is the constraint hyper-surface in Γ and its dimension is less than fourteen by the number of independent primary constraints that eqs. (6.2.16a) imply.

Conservation Laws. The connection between invariances of the action and conservation laws is similar for singular and for non-singular Lagrangians. The invariance of our Lagrangian (6.1.6) under the Poincaré transformation (6.1.1) implies the usual ten conservation laws. Namely the ten phase space variables:

$$M_{\mu\nu} = x_\mu P_\nu - x_\nu P_\mu + b_\mu a_\nu - b_\nu a_\mu ,\ P_\mu \tag{6.2.17}$$

are all constants of motion, and moreover, fulfill via their PB's the Lie algebra relations of the Poincaré group:

$$\begin{aligned} \{M_{\mu\nu}, M_{\rho\sigma}\} &= g_{\mu\rho}M_{\nu\sigma} - g_{\nu\rho}M_{\mu\sigma} + g_{\mu\sigma}M_{\rho\nu} - g_{\nu\sigma}M_{\rho\mu}, \\ \{M_{\mu\nu}, P_\rho\} &= g_{\mu\rho}P_\nu - g_{\nu\rho}P_\mu \quad , \\ \{P_\mu, P_\nu\} &= 0 \quad . \end{aligned} \tag{6.2.18}$$

Thus they generate a realization of the Poincare group by canonical transformations on Γ which are guaranteed to preserve the constraint hypersurface. For later use we recall here the form of the Pauli-Lubanski vector W_μ and its square:

$$W_\mu = \frac{1}{2}\,\varepsilon_{\mu\nu\rho\sigma}P^\nu M^{\rho\sigma} = \varepsilon_{\mu\nu\rho\sigma}P^\nu b^\rho a^\sigma \;,$$

$$W^2 = (P.b)^2 - P^2 b^2 \quad . \qquad (6.2.19)$$

We consider next the three limited models in which the Lagrangian depends only on two of the three variables u_1, u_2, u_3. It turns out possible to give a very complete analysis of each of these. Later on we look briefly at the general case with L involving all three u's.

§3. TWO-VARIABLE MODELS: MODEL A

Let us consider the case when L involves u_1, u_3 alone. It is completely specified by a function f of one argument:

$$L = \sqrt{u_1}\; f(\eta) \quad . \qquad (6.3.1)$$

Denoting the derivative of f with respect to η by a prime, eqs. (6.2.16a) give:

$$P_\mu = \frac{1}{\sqrt{u_1}}(f + 2\eta f')(\dot{x}.a\; a_\mu - x_\mu),$$

$$b_\mu = \frac{2}{\sqrt{u_1}}\,\eta^2 f'\dot{a}_\mu \; . \qquad (6.3.2)$$

We denote the primary constraint $P.a \approx 0$ (present in all the vectorial models) by Φ_2:

$$\Phi_2 = P.a \quad . \qquad (6.3.3)$$

Constraint Corresponding to Chronometric Invariance. Another primary constraint Φ_1 is expected to emerge from the definition of P_μ and b_μ, because of chronometric invariance. Φ_1 must be constructed out of Lorentz scalar combinations of a, P, and b, and the only available ones are P^2, P.b, b^2. From eq. (6.3.2) these turn out to be:

$$\begin{aligned} P^2 &= -(f + 2\eta f')^2 \\ P.b &= -2\eta^2 f'(f + 2\eta f')/\zeta \\ b^2 &= 4\eta^3 f'^2 \quad . \end{aligned} \qquad (6.3.4)$$

Elimination of the Exceptional Cases. The expressions (6.3.2) show that, *in general,* Φ_1 arises by eliminating η between P^2 and b^2, because P.b involves the other variable ζ. In fact, let us define the *generic case* by three conditions: (i) $f'(f + 2\eta f') \neq 0$, so that P.b definitely depends on ζ; (ii) $P^2 \neq$ constant, (iii) $b^2 \neq$ constant. Then Φ_1 is an expression involving P^2 and b^2, whose form is determined by f. If one or more of these three conditions are not obeyed, we define it to be an exceptional case. It is possible to exhibit explicitly all the exceptional cases and eliminate them on physical grounds. They all correspond to f of the form

$$f(\eta) = C_1 + \frac{C_2}{\sqrt{\eta}} ,$$
$$f'(f + 2\eta f') = C_1 C_2 / 2\eta^{3/2}, \quad P^2 = -C_1^2, \; b^2 = C_2^2 , \qquad (6.3.5)$$

where C_1 and C_2 are constants. In fact one finds that if either one of P^2, b^2 is a constant, so is the other. Since in these exceptional cases the invariant mass $\sqrt{P^2}$ is fixed at a constant value, which is uninteresting and unphysical, we will hereafter assume that f is definitely not of the form (6.3.5).

Generic Case: Nontrivial Mass Spectrum. In the generic case the primary constraint Φ_1 expresses P^2 as a nontrivial function of b^2:

$$\Phi_1 = P^2 + \alpha(b^2) = 0. \qquad (6.3.6)$$

The function α is characteristic of the chosen Lagrangian. This will soon be shown to be a relation between the two Casimir invariants P^2, W^2 of the Poincaré group--it is a Regge mass-spin relation.

L being homogeneous of degree one in velocities, the generalized Hamiltonian treatment uses as starting Hamiltonian a linear combination of Φ_1 and Φ_2 with (at this stage) unknown coefficients:

$$H = v_1\Phi_1 + v_2\Phi_2 . \qquad (6.3.7)$$

The equations of motion for the phase space variables are obtained by evaluating the PB of each variable in turn with H and imposing the constraints afterwards:

$$\dot{x}_\mu \approx \{x_\mu, H\} \approx 2v_1P_\mu + v_2a_\mu,\ \dot{a}_\mu \approx \{a_\mu, H\} \approx -2v_1\alpha'(b^2)b_\mu; \quad \text{(a)}$$
$$\dot{P}_\mu \approx \{P_\mu, H\} \approx 0,\ \dot{b}_\mu \approx \{b_\mu, H\} \approx 2v_1\alpha'(b^2)b^2a_\mu + v_2P_\mu . \quad \text{(b)} \qquad (6.3.8)$$

Equations (6.3.8a) are actually the result of "inverting" eq. (6.3.2) to express velocities in terms of momenta: Since P_μ and b_μ are not independent to the extent that their definitions led to two constraints Φ_1, Φ_2, this process of inversion causes the appearance of two arbitrary quantities v_1, v_2. Equations (6.3.8b) are the true Euler-Lagrange equations of motion.

Conditions on the v's and further constraints can arise from the consistency requirement that Φ_1, Φ_2 remain zero always:

$$\dot{\Phi}_1 \approx \{\Phi_1, H\} \approx v_2\{\Phi_1, \Phi_2\} \approx 2v_2\alpha'(b^2)P.b \approx 0,$$
$$\dot{\Phi}_2 \approx \{\Phi_2, H\} \approx -v_1\{\Phi_1, \Phi_2\} \approx -2v_1\alpha'(b^2)P.b \approx 0. \qquad (6.3.9)$$

There are three ways these can be met: (i) both v_1 and v_2 vanish, but that leads to no motion at all; (ii) $\alpha'(b^2)$ vanishes, but that fixes b^2 and hence through (6.3.6) P^2 as well at a constant value independent of the state of motion; (iii) a secondary constraint χ is added to the existing primary ones,

$$\chi = P.b \approx 0. \qquad (6.3.10)$$

We naturally choose this third possibility. (In Lagrangian language this means $u_2 = \dot{x}.\dot{a}$ must vanish.) When we do so, a new consistency condition arises. χ must remain zero always, i.e.,

$$\dot{\chi} \approx \{\chi, H\} \approx v_2\{\chi, \Phi_2\} \approx v_2P^2 \approx 0 \qquad (6.3.11)$$

This forces $v_2 = 0$, since again we do not want to constrain P^2 by itself. The constraint analysis terminates at this stage. The final equations of motion are as in (6.3.8) with the v_2 terms dropped, and the three constraints Φ_1, Φ_2, χ are a self-perpetuating system. Φ_2 and χ form a second class pair, and H is an arbitrary multiple of the sole first class constraint:

$$H = v_1\Phi_1 . \qquad (6.3.12)$$

This is physically very satisfying since it means that the only arbitrary feature characterizing the evolution with respect to s is due to the freedom in choice of s itself--apart from this the motion is fully determined.

Chronometric Invariance Constraint is a Mass-Spin Relation, P^2 *timelike or lightlike.* At this point we can interpret the constraint Φ_1. From eqs. (6.2.9, 6.3.10) we see that in the generic case of model A the square of the Pauli-Lubanski vector is:

$$W^2 \approx -P^2 b^2 \qquad (6.3.13)$$

so relation (6.3.6) expresses *as it stands* the dependence of $(\text{mass})^2$ on "intrinsic spin". Incidentally we also see that b^2 is a constant of motion in this model and that P^2 is timelike or at worst lightlike.

Solution to the Equation of Motion. Let us exhibit the complete solution to the equations of motion--we know that one unknown function, reflecting the arbitrariness in v_1, will be present. Suppose values are chosen at s = 0 for the phase space variables so as to obey all kinematic conditions and constraints:

$$\begin{aligned} &a^2 - 1 = a.b = 0 \ , \\ &P.a \approx P.b \approx P^2 + \alpha(b^2) \approx 0 \end{aligned} \qquad (6.3.14)$$

Then the solution for $x_\mu(s)$, $a_\mu(s)$ and $b_\mu(s)$ is:

$$x_\mu(s) = x_\mu(0) + \frac{\Phi(s)}{\alpha'(b^2)\sqrt{b^2}} P_\mu \ ,$$

$$a_\mu(s) = a_\mu(0) \operatorname{Cos} \Phi(s) - \frac{b_\mu(0)}{\sqrt{b^2}} \operatorname{Sin} \Phi(s) \ ,$$

$$b_\mu(s) = b_\mu(0) \operatorname{Cos} \Phi(s) + a_\mu(0)\sqrt{b^2} \operatorname{Sin} \Phi(s) , \qquad (6.3.15)$$

with $\Phi(s)$ determined by

$$\dot{\Phi}(s) = 2\alpha'(b^2)\sqrt{b^2}\, v_1 , \quad \Phi(0) = 0 \ . \qquad (6.3.16)$$

x^μ traces a straight line (uniformly with respect to $\phi(s)$) parallel to the constant P_μ. The two mutually orthogonal spacelike vectors a_μ, b_μ rotate in the hyperplane orthogonal to P_μ, uniformly and "in phase" with x_μ since the same $\phi(s)$ occurs in all of them. None of the velocities $\dot{x}$, $\dot{x} \pm \dot{a}$ is lightlike in this model.

Dirac Brackets. The PB's (6.2.16b) can be replaced by a system of Dirac brackets (DB) by eliminating the pair of second class constraints Φ_2, χ. Then in any equation stated exclusively in terms of DB's the vanishing of Φ_2 and χ can be treated as identities or strong equations. Since $\{\chi,\Phi_2\} \approx P^2$ the DB between any two phase space functions f, g is

$$\{f,g\}^* \approx \{f,g\} - \frac{1}{P^2}(\{f,P.a\}\{P.b,g\} - \{f,P.b\}\{P.a,g\}). \qquad (6.3.17)$$

The only nonvanishing DB's among x_μ, a_μ, P_μ, b_μ are then:

$$\{x_\mu, P_\nu\}^* \approx g_{\mu\nu},\ \{x_\mu, a_\nu\}^* \approx -a_\mu P_\nu/P^2,\ \{x_\mu, b_\nu\}^* \approx -b_\mu P_\nu/P^2;$$

$$\{x_\mu,x_\nu\}^* \approx (a_\mu b_\nu - a_\nu b_\mu)/P^2 \ ; \qquad (6.3.18)$$

$$\{a_\mu,b_\nu\}^* \approx a_\mu a_\nu + P_\mu P_\nu/P^2 - g_{\mu\nu},\{b_\mu,b_\nu\}^* \approx a_\mu b_\nu - a_\nu b_\mu \ .$$

The general equation of motion is

$$\dot{f} \approx \frac{\partial f}{\partial s} + \{f,\Phi_1\}^* v_1 \qquad (6.3.19)$$

and the weak equality refers to the single constraint $\Phi_1 \approx 0$.

We will leave the description of this model in this manifestly invariant four dimensional form, and not impose yet another constraint (such as say $x^o - s \approx 0$) which fixes v_1 and forms a second class pair (in the DB sense) with Φ_1. The DB's among $M_{\mu\nu}$, P_μ yield again a realization of the Lie algebra of the Poincaré group. This is expected since Φ_2, χ are both Lorentz scalars and have vanishing PB's with $M_{\mu\nu}$, P_μ. The elements of the Poincaré group must now be understood to be implemented by transformations canonical with respect to the DB.

Dirac Bracket of position four vectors is nonzero. It must be noted that the DB $\{x_\mu,x_\nu\}^*$ is nonzero. This causes problems of interpretation for this model when one introduces electromagnetic interactions in the quantized version. It may therefore be useful to explain why this DB has to be nonzero: basically it is because of the momentum dependence in the first constraint we demanded, $P.a \approx 0$. The argument is this: the basic DB's involving P_μ are all fixed by the transformation properties of the variables under space-time translations. Since P.a can be set equal to zero *within* a DB, one finds

$$\{x_\mu, P.a\}^* \approx a_\mu + \{x_\mu, a_\nu\}^* P^\nu \approx 0 . \qquad (6.3.20)$$

This explains the value quoted for $\{x_\mu, a_\nu\}^*$ in eq. (6.3.18). The fact that $\{x_\mu, x_\nu\}^*$ *cannot* vanish is then seen by working out the Jacobi identity

$$\{x_\lambda, \{x_\mu, a_\nu\}^*\}^* + \{x_\mu, \{a_\nu, x_\lambda\}^*\}^* + \{a_\nu, \{x_\lambda, x_\mu\}^*\}^* \approx 0 . \qquad (6.3.21)$$

We close this discussion of the classical aspects of model A in the free case with two comments which show why this particular two-variable model is rather attractive: (i) the vector P_μ in this case is automatically timelike (or lightlike) provided f is real for all physically realized values of η; (ii) there is a clean separation of kinematic and dynamic constraints in this model, in the four-dimensional sense, and for this reason the trajectory function α is absent in the DB's (6.3.18).

§4. TWO VARIABLE MODEL B

Let us take up next the case when L depends on u_1 and u_2 only. The principles of analysis are the same as in the u_3, u_1 case, the details differ. The Lagrangian is determined by a function of ζ:

$$L = \sqrt{u_1}\, f(\zeta); \qquad (6.4.1)$$

and P_μ, b_μ are, from eq. (6.2.16a):

$$P_\mu = \frac{1}{\sqrt{u_1}}(f + 2\zeta f')(\dot{x}.a\, a_\mu - \dot{x}_\mu) - \frac{1}{\sqrt{u_1}}\zeta^2 f \dot{a}_\mu ,$$

$$b_\mu = -\frac{1}{\sqrt{u_1}}\zeta^2 f'(\dot{x}.a\, a_\mu - \dot{x}_\mu) . \qquad (6.4.2)$$

The Lorentz scalars for forming the constraint Φ_1 are

$$P^2 = -f(f + 2\zeta f') + \zeta^4 f'^2/\eta ,$$

$$P.b = \zeta^2 f'(f + \zeta f'),$$

$$b^2 = -\zeta^4 f'^2 . \qquad (6.4.3)$$

Elimination of Exceptional Cases. We are led to define the generic case by the conditions (i) $f \neq 0$, so P^2 definitely involves η, (ii) $P.b \neq$ constant, (iii) $b^2 \neq$ constant. Then Φ_1 arises on eliminating ζ between $P.b$ and b^2. All other cases are exceptional, and they can be quickly disposed of. They all correspond to f of the form

$$f(\zeta) = C_1 + C_2/\zeta \quad ,$$

$$f' = -C_2/\zeta^2, \; P.b = -C_1C_2, \quad b^2 = -C_2{}^2. \tag{6.4.4}$$

In fact if either one of P.b and b^2 is constant, so is the other. In all these cases a straightforward constraint analysis reveals that one never gets an interesting trajectory relation, and the model is physically unsatisfactory. So we now assume f is *not* of the form shown in eq. (6.4.4).

Generic Case: Nontrivial Mass Spectrum from Chronometric Invariance. We will write the constraint Φ_1 due to chronometric invariance as

$$\Phi_1 = P.b - \alpha(b^2) \approx 0 \ , \tag{6.4.5}$$

with the nontrivial function α characterizing the model. The Hamiltonian

$$H = v_1\Phi_1 + v_2\Phi_2 \tag{6.4.6}$$

where the v's are arbitrary and $\Phi_2 = P.a$ as before, generates the equations of motion

$$\dot{x}_\mu \approx v_1 b_\mu + v_2 a_\mu, \quad \dot{a}_\mu \approx v_1(-P_\mu + 2\alpha'(b^2)b_\mu), \qquad \text{(a)}$$
$$\dot{P}_\mu \approx 0, \quad \dot{b}_\mu \approx v_1(P.b - 2\alpha'(b^2)b^2)a_\mu + v_2 P_\mu \ . \qquad \text{(b)} \tag{6.4.7}$$

Equations (6.4.7a) represent the result of inverting eq. (6.4.2) for the velocities. To ensure that Φ_1, Φ_2 remain zero, and discarding the choice $v_1 = v_2 = 0$, we are forced to impose the secondary constraint

$$\chi = P^2 - 2\alpha(b^2)\alpha'(b^2) \approx 0 \ . \tag{6.4.8}$$

(χ is the value of $\{\Phi_1, \Phi_2\}$ modulo the vanishing of Φ_1 itself.) For χ one finds

$$\dot{\chi} = -4\alpha(b^2)(\alpha(b^2)\alpha''(b^2) + \alpha'(b^2))^2 v_2 \ . \tag{6.4.9}$$

We set $v_2 = 0$ to make this vanish: otherwise b^2, and hence by (6.4.8) P^2, would reduce to a constant. The result is that Φ_2, χ form a second class pair, Φ_1 survives as primary first class, and in the equations of motion only one unknown, v_1, is left. One then sees that both P.b and b^2 are constants of motion. The interpretation of Φ_1, or better still χ, as a mass-spin relation is somewhat indirect in this model. From eqs. (6.2.19, 6.4.5, 6.4.8) W^2 is some function of b^2:

$$W^2 \approx \alpha(b^2)^2 - 2b^2\alpha(b^2)\alpha'(b^2) . \qquad (6.4.10)$$

If one imagines solving for b^2 in terms of W^2 and uses this in eq. (6.4.8), there appears a relation between the two Casimir invariants of the Poincaré group.

Solution to the Equation of Motion. For obtaining the complete solution to the equations of motion, it is useful to split b_μ into its components parallel and perpendicular to P :

$$b_{/\!/\mu} = \frac{P.b}{P^2} P_\mu , \quad b_{\perp\mu} = b_\mu - \frac{P.b}{P^2} P_\mu . \qquad (6.4.11)$$

Note that unlike in model A, $b_{/\!/}$ is nonzero now, though it is a constant of the motion; $b_\perp^2$ also is a constant of motion. If we choose values for the variables obeying the kinematic conditions and constraints at s = 0:

$$a^2 - 1 = a.b = 0,$$

$$P.a \approx P.b - \alpha(b^2) \approx P^2 - 2\alpha(b^2)\alpha'(b^2) \approx 0 \qquad (6.4.12)$$

then the solution for x_μ, a_μ and $b_{\perp\mu}$ is

$$x_\mu(s) = x_\mu(0) + \frac{\Phi(s)}{4\sqrt{b_\perp^2}(\alpha'(b^2))^2} P_\mu + \frac{a_\mu(s) - a_\mu(0)}{2\alpha'(b^2)},$$

$$a_\mu(s) = a_\mu(0) \operatorname{Cos}\Phi(s) + \frac{b_{\perp\mu}(0)}{\sqrt{b_\perp^2}} \operatorname{Sin}\Phi(s),$$

$$b_{\perp\mu}(s) = b_{\perp\mu}(0) \operatorname{Cos}\Phi(s) - a_\mu(0)\sqrt{b_\perp^2} \operatorname{Sin}\Phi(s). \qquad (6.4.13)$$

$\Phi(s)$ is to be found by solving

$$\dot\Phi(s) = 2\sqrt{b_\perp^2}\,\alpha'(b^2)v_1, \quad \Phi(0) = 0. \qquad (6.4.14)$$

x_μ undergoes, on the average, uniform translation parallel to P_μ , but superimposed on this is a rotating component in the hyperplane orthogonal to P_μ. The two vectors a_μ and $b_{\perp\mu}$, perpendicular to one another and to P_μ, rotate into one another, uniformly and in step with x_μ.

While this model is qualitatively quite satisfactory, it is somewhat less attractive than model A for two reasons: (i) the momentum P_μ is not guaranteed anymore to be timelike, thus reflecting a kinematic difference between the pair u_1, u_3 and the pair u_1, u_2; (ii) the "trajectory function" α occurs both in the second class pair Φ_2, χ and in Φ_1. For the latter reason the DB's that we get on eliminating Φ_2 and χ are not at all simple and "kinematic" in appearance. We therefore do not pursue the analysis of this model further.

§5. TWO VARIABLE MODEL C

The last of the two-variable models corresponds to a Lagrangian formed out of u_2 and u_3. Let us write

$$\rho = \frac{\eta}{\zeta} = \frac{u_2}{u_3} \tag{6.5.1}$$

Then the model is determined by a function $g(\rho)$,

$$L = \sqrt{u_2}\; g(\rho) \tag{6.5.2}$$

and this corresponds to the function f of eq. (6.1.6) being

$$f(\zeta,\eta) = \frac{1}{\sqrt{\zeta}}\; g\;(\rho) \tag{6.5.3}$$

Actually this model does not lead to a physically interesting system, and this reflects the properties of the pair u_2, u_3. The variable u_1 represents a minimal change in $\dot{x}^2$, which one uses for the point particle, to accommodate the requirement $P.a \approx 0$; so in both models A and B we obtained physically reasonable results. This is missing in the present case, but for completeness we show briefly what happens.

Using eq. (6.5.3) in eq. (6.2.16a) we get P_μ and b_μ:

$$P_\mu = \frac{1}{\sqrt{u_2}}(\rho g' + \frac{g}{2})\; \dot{a}_\mu \quad ,$$

$$b_\mu = \frac{1}{\sqrt{u_2}}\; (\rho g' + \frac{g}{2})\,(\dot{x}.a\; a_\mu - \dot{x}_\mu) + \frac{2}{\sqrt{u_2}}\; \rho^2 g' \dot{a}_\mu\,. \tag{6.5.4}$$

Then we compute the combinations P^2, P.b, b^2:

$$P^2 = \frac{1}{\rho}\,(\rho g' + \frac{g}{2})^2 \ ,$$

$$P.b = (\rho g')^2 - \frac{g^2}{4} \ , \qquad (6.5.5)$$

$$b^2 = -2\rho^2 g g' - (\rho g' + \frac{g}{2})^2 \zeta$$

One defines the generic case as obtaining when (i) $\rho g' + \frac{g}{2} \neq 0$, so that b^2 definitely depends on ζ, (ii) $P^2 \neq$ constant, (iii) $P.b \neq$ constant. In this case, the primary constraint Φ_1 relates P^2 and P.b. If one or more of the three conditions fail, we have an exceptional case. All such correspond to g of the form

$$g(\rho) = C_1\sqrt{\rho} + C_2/\sqrt{\rho} \ ,$$
$$(6.5.6)$$
$$\rho g' + \frac{g}{2} = C_1\sqrt{\rho} \ , \quad P^2 = C_1{}^2, \quad P.b = -C_1C_2 \ ,$$

and can immediately be discarded. In the generic case we express Φ_1 as

$$\Phi_1 = P^2 + \alpha(P.b) \approx 0. \qquad (6.5.7)$$

The constraint analysis starts with the Hamiltonian

$$H = v_1\Phi_1 + v_2\Phi_2 \ . \qquad (6.5.8)$$

(Φ_2 = P.a as before). The equations of motion for Φ_1 and Φ_2 are:

$$\dot{\Phi}_1 \approx \{\Phi_1,\Phi_2\}v_2 \approx v_2\alpha'(P.b)P^2, \quad \dot{\Phi}_2 \approx \{\Phi_2,\Phi_1\}v_1 \approx -v_1\alpha'(P.b)P^2 \qquad (6.5.9)$$

These must vanish. If we avoid setting $v_1 = v_2 = 0$, we have no alternative but to choose as secondary constraint either $P^2 \approx 0$ or $\alpha'(P.b) \approx 0$. Because of the already existing primary constraint Φ_1, in either case we end up with P^2 being a constant. Thus even in the generic case model C does not yield an interesting physical system.

In concluding our discussion of the classical features of the vectorial two-variable models we point out an interesting correspondence between the Lagrangian variables u_1, u_2, u_3 and the phase space combinations P^2, P.b, b^2:

$$u_1 \leftrightarrow b^2, \quad u_2 \leftrightarrow P.b, \quad u_3 \leftrightarrow P^2. \tag{6.5.10}$$

The sense of this correspondence is that in the generic case of a two-variable model with L a function of u_j and u_k, only the two phase space Lorentz scalars corresponding to u_j and u_k occur in the primary constraint Φ_1 that is caused by the chronometric invariance.

§6. THE GENERAL THREE-VARIABLE CASE

It is difficult to give a complete analysis and classification in this case, into generic and exceptional Lagrangians, since we have to deal with a function of two independent arguments. However, the overall structure of the model can be exhibited fairly easily. Let us write the expressions for P_μ and b_μ as

$$P_\mu = \frac{1}{\sqrt{u_1}} (A\,(\dot{x}.a\; a_\mu - \dot{x}_\mu) + B\mathring{a}_\mu),$$

$$b_\mu = \frac{1}{\sqrt{u_1}} (B\,(\dot{x}.a\; a_\mu - \dot{x}_\mu) + C\mathring{a}_\mu), \tag{6.6.1}$$

where A, B, C are functions of ζ, η that can be identified from eq. (6.2.16a)

$$A = f + 2\zeta \frac{\partial f}{\partial \zeta} + 2\eta \frac{\partial f}{\partial \zeta}, \quad B = \zeta^2 \frac{\partial f}{\partial \zeta},$$

$$C = 2\eta^2 \frac{\partial f}{\partial \eta}. \tag{6.6.2}$$

Constraint Corresponding to Chronometric Invariance. The expressions that will lead to the primary constraint Φ_1 are:

$$P^2 = -A^2 - 2\frac{AB}{\zeta} + \frac{B^2}{\eta},$$

$$P.b = -AB - \frac{(AC + B^2)}{\zeta} + \frac{BC}{\eta},$$

$$b^2 = -B^2 - \frac{2BC}{\zeta} + \frac{C^2}{\eta}. \tag{6.6.3}$$

Let us assume for definiteness that none of these three expressions reduces to a constant. We then produce Φ_1 by eliminating ζ and η between them, and write it in the form

$$\Phi_1 = P^2 + \alpha(P.b,\, b^2) \approx 0 . \tag{6.6.4}$$

There is thus a correspondence between the function f in the Lagrangian, dependent on two arguments, and the function α here, again dependent on two arguments.

Equations of Motion. Taking as Hamiltonian a combination of Φ_1, Φ_2 with arbitrary coefficients v_1, v_2, we get equations of motion

$$\begin{aligned} \dot{x}_\mu &\approx v_1 \, (2P_\mu + \alpha_1 b_\mu) + v_2 a_\mu \ , \\ \dot{a}_\mu &\approx -v_1 (\alpha_1 P_\mu + 2\alpha_2 b_\mu) , \\ \dot{P}_\mu &\approx 0 , \\ \dot{b}_\mu &\approx v_1 (\alpha_1 P.b + 2\alpha_2 b^2) a_\mu + v_2 P_\mu \ . \end{aligned} \tag{6.6.5}$$

Here we have written α_1, α_2 for the derivatives of α with respect to its first and second arguments respectively. Then since

$$\begin{aligned} \dot{\Phi}_1 &\approx (\alpha_1 P^2 + 2\alpha_2 P.b) v_2 \ , \\ \dot{\Phi}_2 &\approx - (\alpha_1 P^2 + 2\alpha_2 P.b) v_1 \ , \end{aligned} \tag{6.6.6}$$

we must impose the secondary constraint

$$\chi = \alpha_1 P^2 + 2\alpha_2 P.b \approx 0 \ . \tag{6.6.7}$$

At this point we see that of the three variables P^2, P.b, b^2 only one is independent since both Φ_1 and χ must vanish. Turning to the maintenance of the χ constraint, we find that $\{\chi, \Phi_1\}$ vanishes (weakly) and, viewing χ as an expression in P^2, P.b and b^2,

$$\dot{\chi} \approx \{\chi, P.a\} v_2 \approx (P^2 \frac{\partial\chi}{\partial P.b} + 2P.b \frac{\partial\chi}{\partial b^2}) \ v_2 \ . \tag{6.6.8}$$

If we did not set $v_2 = 0$, we would produce a third constraint on P^2, P.b, b^2, and that would reduce them all to constants. To avoid this we put $v_2 = 0$. At the end of the constraint analysis Φ_1 is first class, Φ_2 and χ second class, and v_1 is the only unknown in the dynamical equations. Thus these features carry over from the two-variable models A and B to the general case.

Solution to the Equations of Motion. One can solve the equations of motion (6.6.5) explicitly again, since P.b and b^2, and so α_1, α_2 as well, are constants of motion. In this way we obtain:

$$x_\mu(s) = x_\mu(0) + \frac{(4\alpha_2 - \alpha_1{}^2)}{2\alpha_2\Lambda}\,\Phi(s)P_\mu - \frac{\alpha_1}{2\alpha_2}\,(a_\mu(s) - a_\mu(0)),$$

$$a_\mu(s) = a_\mu(0)\ \mathrm{Cos}\,\Phi\,(s) - \frac{2\alpha_2 b_{\perp\mu}{}^{(0)}}{\Lambda}\quad \mathrm{Sin}\,\Phi\,(s),$$

$$b_{\perp\mu}(s) = b_{\perp\mu}(0)\ \mathrm{Cos}\,\Phi\,(s) + \frac{\Lambda}{2\alpha_2}\,a_\mu(0)\ \mathrm{Sin}\,\Phi\,(s), \qquad (6.6.9)$$

where

$$\dot{\phi}(s) = \Lambda v_1, \quad \phi(0) = 0,$$

$$\Lambda = \sqrt{2\alpha_2(\alpha_1 P.b + 2\alpha_2 b^2)} \qquad (6.6.10)$$

The qualitative features of the motion are similar to model B.

Dirac Brackets $\{x_\mu, x_\nu\}^*$ *do not commute.* If we pass to a system of DB's to eliminate the second class pair Φ_2, χ, we find the basic brackets depend on the function α, and also $\{x_\mu, x_\nu\}^*$ does not vanish.

§7. QUANTIZATION

We shall discuss this only in the context of the two-variable model A for reasons already alluded to: this is the model with the simplest constraint structure, and at the classical level the energy-momenta occurring in the theory are guaranteed to be timelike or lightlike.

Conventional Way to Proceed. The conventional way to quantize the classical theory actually destroys the manifest relativistic invariance we have carefully maintained (though hopefully relativistic invariance itself is saved). Namely one lifts the freedom in the choice of s, reflected in the arbitrariness in v_1 in the final equation of motion (6.3.17), by adopting a so-called gauge constraint χ' which has explicit s dependence. Its purpose is to convert one of the phase space dynamical variables into the parameter s and to achieve this χ' and Φ_1 must form a second class pair with respect to the DB, i.e. $\{\chi', \Phi_1\}^{*1}$ must be non-zero. A common choice is $\chi' = x^0 - s$. The pair χ', Φ_1 can now be eliminated, that is these two constraints can be made strong equations by passing to a final system of DB's $\{\ ,\ \}^{**}$ using the iterative property of such brackets. In this final form of the classical theory, only the right number of independent dynamical variables remains: they are $\vec{x}, \vec{P}, \hat{a}$ and $\hat{b}$. The final DB's among these will, however, in general involve the trajectory function α of the model. The maintenance of relativistic invariance at the classical level is guaranteed by this fact: with respect to the final DB's the ten quantities $M_{\mu\nu}$, P_μ do obey the Lie relations of the Poincaré group. Having put the classical theory into this form which is relativistically invariant but not manifestly so, one can try to set up a quantum theory by making $\vec{x}$, $\vec{P}$, $\hat{a}$, $\hat{b}$ into hermitian operators and converting the basic $\{\ ,\ \}^{**}$ brackets into commutation relations. The problem of ordering non-commuting factors can arise in stating the commutation relations as well as in defining $M_{\mu\nu}$, P_μ as hermitian operators. If one succeeds in finding a representation of the commutation relations in a suitable Hilbert space and moreover $M_{\mu\nu}$, P_μ give a hermitian operator representation of the Lie algebra of the Poincaré group, one has an acceptable relativistic quantum theory based on the classical model. The Hilbert space and its inner product will be the physical ones appropriate for the usual quantum mechanical interpretation, and the operator P_o provides us with the Schrödinger equation (provided $x' = x^o - s$ was chosen!).

Heuristic Covariant Procedure. We will not follow the above procedure because of the α dependence of the brackets $\{\ ,\ \}^{**}$, but will adopt a somewhat heuristic procedure which maintains the manifest relativistic invariance throughout. The idea is to convert the classical constraint $\Phi_1 \approx 0$ into a wave equation on a suitable wave function ψ, and view it as a generalization of the Klein-Gordon equation with a mass operator $\alpha(b^2)$ in place of a numerical mass parameter. To do this we must make all components of the four-vectors x_μ, a_μ, P_μ, b_μ into operators on a suitable Hilbert space, obeying commutation relations suggested by the DB's (6.3.18). We will see that we can consistently arrange to have all these operators hermitian. Now this Hilbert space and its inner product are purely mathematical, and are not to be used for the normal quantum-mechanical interpretation. One may then ask: why should one look for hermitian operators x,a,P,b at all at this level? The answer is that one would like to have a mass operator $\alpha(b^2)$ which is hermitian in a suitable space and so we are sure to end up with real mass values in the solutions of the wave equation. The situation is similar to viewing the ordinary Klein-Gordon equation

$$(\Box^2 - m^2)\ \phi(x) = 0 \tag{6.7.1}$$

as arising from the classical mass shell constraint relation

$$P^2 + m^2 \approx 0 \tag{6.7.2}$$

by the formal replacement $P_\mu \rightarrow -i\dfrac{\partial}{\partial x^\mu}$ in the constraint and then applying the resulting operator to a complex function $\phi(x)$ on space-time. The four operators $-\dfrac{i\partial}{\partial x^\mu}$ are hermitian in the mathematical Hilbert space of all functions $\phi(x)$ that are square integrable over all space-time, but the physical Hilbert space and inner product are to be formed only within the space of (positive energy) solutions of the Klein-Gordon equation. The situation is qualitatively similar in our vectorial model.

At this point we must mention one important difference between the Klein-Gordon equation and our model: on quantization the latter will not have a direct space-time visualizability in any manifestly covariant sense at all. This is because the only way to quantize a classical constrained theory is to first eliminate all second class constraints and then convert the resulting DB's into commutators. One cannot carry second class constraints into quantum theory as operator conditions on a wave function. But the DB's $\{x_\mu, x_\nu\}^*$ in eq. (6.3.18) are non-zero. So in the quantum theory we cannot have a representation with the x_μ simultaneously diagonal--we will not find wave functions defined on space-time, and the internal space, at all. The construction has to be in momentum space entirely.

Commutation Relations. After this preamble let us proceed to the details. We begin by postulating a set of commutation relations based on the DB's (6.3.18); the complete set is written in a natural sequence as

$$[P_\mu, P_\nu] = [P_\mu, a_\nu] = [a_\mu, a_\nu] = 0; \qquad \text{(a)}$$

$$[x_\mu, P_\nu] = i\hbar g_{\mu\nu},\ [x_\mu, a_\nu] = -i\hbar a_\mu P_\nu / P^2,$$

$$[b_\mu, P_\nu] = 0, \quad [b_\mu, a_\nu] = i\hbar\,(g_{\mu\nu} - a_\mu a_\nu - P_\mu P_\nu/P^2); \qquad \text{(b)}$$

$$[x_\mu, x_\nu] = i\hbar\,(a_\mu b_\nu - a_\nu b_\mu)/P^2,\ [x_\mu, b_\nu] = i\hbar b_\mu P_\nu/P^2, \qquad (6.7.3)$$

$$[b_\mu, b_\nu] = i\hbar\,(a_\mu b_\nu - a_\nu b_\mu). \qquad \text{(c)}$$

One can check that (i) all the Jacobi identities are obeyed, (ii) it is consistent to demand that x, a, P, b be hermitian, (iii) it is also consistent to demand that $a^2 - 1$, $P.a$, $P.b$ vanish as operators. Due to hermiticity and the [a,b] commutator, one cannot demand that a.b also vanish; but we shall see later that the symmetrized hermitian combination $a.b + b.a$ does vanish.

Realization on a Hilbert Space. We now want to find a hermitian irreducible solution to (6.7.3) in a suitable mathematical Hilbert space $\mathcal{H}_M$. From (6.7.3a) one may assume that P_μ and a_μ are simultaneously diagonal. We denote eigenvalues of these operators by the same symbols, for simplicity. A general element of $\mathcal{H}_M$ will then be a complex valued function $\psi(P,a)$ of two real four-vectors subject to $a^2 - 1 = P.a = 0$. Though ψ is thus a function on a six dimensional manifold we will use both P_μ and a_μ as arguments to maintain a manifestly relativistic appearance. We define the norm in $\mathcal{H}_M$ by

$$\| \psi \|_M^2 = \int d\mu(P,a) \; |\psi(P,a)|^2 \quad ,$$

$$d\mu(P,a) = d^4P \, d^4a \, \delta(a^2 - 1)\delta(P.a) \quad . \tag{6.7.4}$$

The next set of commutation relations (6.7.3b) tells us how x_μ and b_μ alter the arguments when applied to a $\psi(P.a)$. For an infinitesimal numerical four vector ε_μ define the operators

$$U(\varepsilon) \simeq 1 + i\varepsilon.x \; , \quad V(\varepsilon) \simeq 1 + i\varepsilon.b. \tag{6.7.5}$$

If these are infinitesimal unitary operators then x_μ and b_μ will be hermitian. (ε is real!) The commutation relations (6.7.3b) take the suggestive form

$$U(\varepsilon)P_\mu U(\varepsilon)^{-1} = P_\mu - \hbar\varepsilon_\mu \; , \quad U(\varepsilon)a_\mu U(\varepsilon)^{-1} = a_\mu + \hbar\varepsilon.aP_\mu/P^2 ,$$
$$V(\varepsilon)P_\mu V(\varepsilon)^{-1} = P_\mu \quad , \quad V(\varepsilon)a_\mu V(\varepsilon)^{-1} = a_\mu - \hbar\varepsilon_\mu + h\varepsilon.a\, a_\mu + \hbar\varepsilon.P\, P_\mu/P^2 . \tag{6.7.6}$$

We infer that the effects of $U(\varepsilon)$, $V(\varepsilon)$ on a function $\psi(P,a)$ must be

$$(U(\varepsilon)\psi)(P,a) = \lambda(P,a;\varepsilon)\psi\,(P-\hbar\varepsilon,\; a + \hbar\varepsilon.a\; P/P^2\,) ,$$
$$(V(\varepsilon)\psi)(P,a) = \nu(P,a;\varepsilon)\psi(P,a - \hbar\varepsilon + \hbar\varepsilon.a\; a + \hbar\varepsilon.P\; P/P^2) , \tag{6.7.7}$$

where λ, ν are two multipliers left undetermined by (6.7.6) but restricted by unitarity. The arguments of ψ on the right hand side in (6.7.7) obey the same restrictions as those on the left, so both U and

V map a ψ in H_M into H_M. To get the consequences of unitarity on λ and ν, we see how the measure $d_\mu(P,a)$ alters under the changes of argument occurring in (6.7.7). For the case of $U(\varepsilon)$, if we write $P' = p - \hbar\varepsilon$, $a' = a + h\varepsilon.a\, P/P^2$, we see that

$$d\mu(P',a') = d^4P'd^4a'\delta(P'a')\delta(a'^2 - 1) \qquad (6.7.8)$$
$$= (1 + \hbar\varepsilon.P/P^2)\, d\mu(P,a).$$

So $U(\varepsilon)$ will be unitary on H_M, and x_μ hermitian, if

$$|\lambda(P,a;\varepsilon)|^2 \simeq 1 + \hbar\varepsilon.P/P^2. \qquad (6.7.9)$$

We shall choose λ to be real and positive,

$$\lambda(P,a;\varepsilon) \simeq 1 + \hbar\varepsilon.P/2P^2, \qquad (6.7.10)$$

and so obtain the following form for x_μ:

$$x_\mu = -\frac{i\hbar}{2}P_\mu/P^2 + i\hbar\left(\frac{\partial}{\partial P^\mu} - \frac{a_\mu}{P^2}P\cdot\frac{\partial}{\partial a}\right). \qquad (6.7.11)$$

In a similar way we use unitarity of $V(\varepsilon)$ to make the choice

$$\nu(P,a;\varepsilon) \simeq 1 + \hbar\, a.\varepsilon \qquad (6.7.12)$$

and arrive at the following form for b_μ:

$$b_\mu = i\hbar a_\mu + i\hbar\left(\frac{\partial}{\partial a^\mu} - a_\mu a.\frac{\partial}{\partial a} - \frac{P_\mu}{P^2}P\cdot\frac{\partial}{\partial a}\right). \qquad (6.7.13)$$

Thus far then the commutation relations (6.7.3a,b) have been solved on H_M with hermitian operators.

The commutation relations (6.7.3c) bilinear in x and b, now remain. It is possible to check that the operators (6.7.11, 6.7.13) in fact obey these relations. For this purpose it is useful to imagine that any $\psi(P,a)$ is defined even away from the region $P.a = a^2 - 1 = 0$, and one may treat all eight components P_μ, a_μ as independent variables for applying the differential operators (6.7.11, 6.7.13) to ψ: it is the case that the specific combinations of $\frac{\partial}{\partial P^\mu}$ and $\frac{\partial}{\partial a^\mu}$ occurring in x and do not disturb the conditions $P.a = a^2 - 1 = 0$. Keeping this remark in mind, the validity of (6.7.3c) follows.

We see that a.b does not vanish but has the value $-i\hbar$. This is consistent with a_μ and b_μ being hermitian, and with the result $a.b - b.a = 2i\hbar$ implied by the $[a_\mu, b_\nu]$ commutator. Having obtained the important expressions (6.7.11, 6.7.13) for x_μ and b_μ, we now discard the Hilbert space H_M as being of no physical significance. At any rate we now define physical quantum mechanical wave functions $\psi(P,a)$ to be solutions of the wave equation.

$$(P^2 + \alpha(b^2))\ \psi(P,a) = 0. \tag{6.7.14}$$

We will assume that α is a real positive definite function of its argument, as is natural due to the classical formula (6.3.6) for P^2. In this wave equation P^2 is a number, and b^2 is an operator acting on the argument a_μ in ψ. Of course again ψ is defined only over configuration obeying $P.a = a^2 - 1 = 0$.

It is clear physically that $\psi(P,a)$ vanishes unless P is timelike: this is the consequence of the positive definiteness of $\alpha(b^2)$. One can find the most general solution to eq. (6.7.14) as follows. Since P_μ has to be timelike, we can use the pure Lorentz transformation $\Lambda(P)$ of eq. (2.4.2) to simplify the problem. We extend eq. (2.4.2) to allow both signs for P_o and define:

$$\Lambda(P)^\mu{}_o = \Lambda(P)^o{}_\mu = \varepsilon\frac{P^\mu}{M},$$

$$\Lambda(P)^j{}_k = \delta_{jk} + \frac{P_j P_k}{\vec{p}\cdot\vec{p}}\left(\varepsilon\frac{P^0}{M} - 1\right) \tag{6.7.15}$$

$$\varepsilon = \text{sign } P^0, \quad M = +\sqrt{-P^2}$$

Then we pass from the pair of variables P,a to P,a' where

$$a'^\mu = \Lambda(P)^\mu{}_\nu a^\nu . \tag{6.7.16}$$

The wave function $\psi(P,a)$ can be rewritten as some function of P and a':

$$\psi(P,a) = \phi(P,a'). \tag{6.7.17}$$

Because $\Lambda(P)$ is the Lorentz transformation that relates P_μ to its rest frame form, the restrictions $P.a = a^2 - 1 = 0$ on the arguments of ψ imply

$$a'^o = 0, \quad |\vec{a}'| = 1 \tag{6.7.18}$$

for the arguments of ϕ. Thus ϕ is a wave function dependent on a timelike P_μ and a purely spatial unit vector $\hat{a}'$, which can be treated as completely independent variables. The operator b_μ can be converted from the form (6.7.13) to one in which partial derivatives with respect to $\hat{a}'$ appear, so that it can be applied to a $\phi(P,\hat{a}')$. Setting $\hbar = 1$, a little bit of algebra shows that

$$b^2 = 1 + \vec{L}'\cdot\vec{L}', \quad \vec{L}' = -i\hat{a}'\times\nabla_{\hat{a}'} \tag{6.7.19}$$

(Thus $\vec{L}'$ is the quantum mechanical orbital angular momentum operator corresponding to $\hat{a}'$). By expanding $\phi(P,\hat{a}')$ in a series of spherical harmonics

$$\phi(P,\hat{a}') = \sum_{\ell m} \phi_{\ell m}(P)\, Y_{\ell m}(\hat{a}'), \tag{6.7.20}$$

and writing for simplicity

$$\alpha(\ell(\ell+1)+1) = M(\ell)^2 , \tag{6.7.21}$$

the wave eq. (6.7.14) separates into independent equations on $\phi_{\ell m}$:

$$(P^2 + M(\ell)^2)\,\phi_{\ell m}(P) = 0 . \tag{6.7.22}$$

Similar to the Klein-Gordon case, positive and negative energy solutions occur symmetrically, and the amplitude $\phi_{\ell m}(P)$ "lives" only on the mass hyperboloid $P^2 = -M(\ell)^2$. One may write the most general solution as

$$\phi(P,\hat{a}') = \sum_{\ell m} \Big[c_{\ell m}(\vec{P})\, \delta(P^0 - \sqrt{\vec{P}^2 + M(\ell)^2}) + d_{\ell m}(\vec{P})\, \delta(P^0 + \sqrt{\vec{P}^2 + M(\ell)^2}) \Big] \frac{Y_{\ell m}(\hat{a}')}{\sqrt{\vec{P}^2 + M(\ell)^2}} , \tag{6.7.23}$$

and recover the most general ψ with the help of eq. (6.7.16).

If we restrict ourselves to the positive energy solutions ($d_{\ell m} = 0$) it is obvious that basically we have a direct sum of time-like unitary irreducible representations of the Poincare group, each integer spin representation $\ell = 0,1,2,$ carrying mass $M(\ell)$. The problem of setting up a physical Hilbert space structure (which we do not solve completely) remains in this guise: it is clear that we must define the physical norm of a solution $\psi(P,a)$ through something like

$$||\psi||^2_{phys} = \sum_{\ell=0,1,\ldots} \mu_\ell \int \frac{d^3P}{\sqrt{\vec{P}^2 + M(\ell)^2}} \sum_m |c_{\ell m}(\vec{P})|^2 \tag{6.7.24}$$

where there is some freedom in the choice of the real positive constants μ_ℓ. For the free system the choice of the μ_ℓ is not too crucial, but they do affect the way in which x_μ and b_μ, hitherto well-defined operators on H_M, can be turned into hermitian operators on the space of solutions of the wave equation. This is a nontrivial problem but we do not pursue it since, as we shall see presently, there is no consistent way of introducing interactions in the vectorial model.

§8. ELECTROMAGNETIC INTERACTION

Let us return to the classical theory and to the Lagrangian (6.3.1), where it is assumed that f is generic. At this level it is easy enough to introduce interactions with an external electromagnetic field with vector potential $A_\mu(x)$ by adding an interaction term

$$L_I = e\, A_\mu(x)\, \dot{x}^\mu \ . \tag{6.8.1}$$

Notice that L_I is homogeneous of degree one in the velocities, so chronometric invariance is maintained. We also have gauge invariance: the replacement $A_\mu(x) \to A_\mu(x) + \partial_\mu \Lambda(x)$ changes the Lagrangian by $\dot{\Lambda}$. In computing the canonical momenta P_μ, b_μ we see that only the former changes:

$$\begin{aligned} P_\mu - eA_\mu \equiv \Pi_\mu &= \frac{1}{\sqrt{u_1}}(f + 2\eta f')(\dot{x}.a\ a_\mu - \dot{x}_\mu), \\ b_\mu &= \frac{2}{\sqrt{u_1}}\ \eta^2 f' \dot{a}_\mu \ . \end{aligned} \tag{6.8.2}$$

Thus the combinations Π^2, $\Pi.b$, b^2 are the *same* functions of ξ and η (for given f) as P^2, $P.b$, b^2 were in the free case. Since f is generic, the two primary constraints now are

$$\Phi_1 = \Pi^2 + \alpha(b^2) \approx 0, \qquad \Phi_2 = \Pi.a \approx 0. \tag{6.8.3}$$

Here α is the "free" trajectory function; and one can proceed with the constraint analysis in the familiar way. But it is clear that the final DB's for the system, which will now be field dependent, will entail nonzero value for $\{x_\mu, x_\nu\}^*$. *Thus though there are no problems of principle in introducing electromagnetic interactions in the classical vectorial model, the same problems of factor ordering and of interpretation occur here as in the relativistic top model of Hanson and Regge.* In fact our earlier discussion at the end of §3 shows that such problems are expected to be present in any theory with momentum dependent second class constraints.

CHAPTER SEVEN

LAGRANGIAN SPINOR MODEL, ELECTROMAGNETIC INTERACTION

We take up now the construction of a classical Lagrangian theory for an indecomposable object whose internal space is spanned by spinor variables. There are two reasons to do this. The first reason is that in chapters 3 and 4 we formed a quantum mechanical model of this type. The model described a Regge sequence of free particles and one would like to extend it so that interactions, in particular with the electromagnetic field, can be introduced. The second reason is also related to the introduction of interaction with an electromagnetic field: the spinor model is more convenient than the vector model of chapter 6 for introducing such interactions. The models of chapter 6, even in the manifestly covariant form, have second class constraints. These second class constraints make the Dirac brackets $(x_\nu, x_\nu)^*$ nonzero, see (6.3.18), a feature which the models of chapter 6 share with the model of Regge and Hanson, discussed in Appendix A. The Dirac brackets of x_μ and x_ν being nonzero is quite all right for the classical theory, but it introduced difficulties on quantizing (replacing the brackets by commutators) as then, the components of the four vector x^μ in $A_\mu(x)$ and $F_{\mu\nu}(x)$ do not commute. The spinor model of the present chapter does not have these difficulties, in the "manifestly covariant" form of the model there is only one constraint, which therefore is first class. Thus the Dirac brackets, at that stage, are the original Poisson brackets with $\{x_\mu, x_\mu\}^* = 0$. Thus the simpler constraint algebra is of great importance in introducing external electromagnetic fields for the quantum version of this model. [MUK 3].

§1. CHOICE OF VARIABLES AND LAGRANGIAN

Basic Variables. The basic variables will be a space-time position four vector x^μ and a real four-component spinor $Q_a = (\xi_1, \xi_2, \pi_1, \pi_2)$. The latter is intended to be the classical limit of the set of internal operators used in chapter 2 to set up the new Dirac equation. Thus for the present all the Q_a will be real classical variables. Under a general element (Λ, d) in the Poincaré group, we make x and Q transform as

$$(\Lambda, d): \quad x^\mu \to x'^\mu = \Lambda^\mu{}_\nu x^\nu + d^\mu ,$$

$$Q \to Q' = S(\Lambda)\, Q. \qquad (7.1.1)$$

$S(\Lambda)$ is the 4 x 4 matrix defined in Chapter 2, §2 and forms the real 4 x 4 spinor representation of SL(2,C). When Ω transforms as above, eqs. (2.2.27) show that the classical bilinear expressions

$$S_{\mu\nu} = -\frac{1}{8}\,\Omega^{\sim}\beta^{-1}[\gamma_\mu,\gamma_\nu]\Omega, \quad V_\mu = \frac{1}{4}\Omega^{\sim}\beta^{-1}\gamma_\mu\Omega \tag{7.1.2}$$

transform respectively as an antisymmetric tensor and a vector. (The matrix algebra of chapter 2 is used freely here). These classical quantities obey identities which are the limiting forms as $h \to 0$ of the operator identities given in eq. (2.2.25). Thus we now have

$$\begin{aligned} &S_{\mu\nu}S^{\mu\nu} = \varepsilon^{\nu\rho\sigma}S_{\mu\nu}S_{\rho\sigma} = V_\mu V^\mu = 0, \\ &S_{\mu\nu}V^\nu = \varepsilon_{\mu\nu\rho\sigma}S^{\nu\rho}V^\sigma = 0, \\ &S_{\lambda\mu}S^\lambda{}_\nu = V_\mu V_\nu \quad . \end{aligned} \tag{7.1.3}$$

To verify these we may use the expressions given in eqs. (2.2.17).

We suppose x^μ and Ω are functions of an evolution parameter s, so that $x^\mu(s)$ traces a world line in space time for our object. In contrast to the vectorial model, now the internal variable $\Omega(s)$ cannot be directly visualized in a space-time diagram, though the quadratic quantities $S_{\mu\nu}(s)$, $V_\mu(s)$ can. This, however, does not prevent our using Ω as a dynamical variable in a Lagrangian, provided we apply the normal procedures of the canonical formalism to it as to x^μ.

Lagrangian. We now look for the most general Lagrangian L formed out of $x, \dot{x}, \Omega, \dot{\Omega}$, with these three properties: (i) on application of the canonical formalism it must turn out that ξ_1, Π_1 and ξ_2, Π_2 form two independent canonical pairs; (ii) L must be invariant under the action (7.1.1) of the Poincaré group; (iii) it must be chronometrically invariant.

Let us examine these conditions in term. Condition (i) says there should be no need to introduce four new quantities to serve as canonical conjugate to Ω_a, but that Ω contains within itself complete canonical pairs. This will happen if L depends on Ω in a special way, namely

$$\begin{aligned} L &= \frac{1}{2}(\pi_1\dot{\xi}_1 - \dot{\pi}_1\xi_1 + \pi_2\dot{\xi}_2 - \dot{\pi}_2\xi_2) + L'(x,\dot{x},\Omega) \\ &= \frac{1}{2}\dot{\Omega}^{\sim}\beta\Omega + L'(x,\dot{x},\Omega). \end{aligned} \tag{7.1.4}$$

With respect to Q, L is singular in the Dirac sense, and if one goes through the usual procedures one indeed gets in analogy with example 1 of chapter 5 the result that among the Q_a the following PB relations hold:

$$\{Q_a, Q_b\} = \beta_{ab} \tag{7.1.5}$$

With these PB's one recovers the classical version of the various commutation properties among $S_{\mu\nu}$, V_μ, Q in chapter 2. Namely one has the PB relations

$$\{S_{\mu\nu}, S_{\rho\sigma}\} = g_{\mu\rho}S_{\nu\sigma} - g_{\nu\rho}S_{\mu\sigma} + g_{\mu\sigma}S_{\rho\nu} - g_{\nu\sigma}S_{\rho\mu},$$

$$\{S_{\mu\nu}, V_\rho\} = g_{\mu\rho}V_\nu - g_{\nu\rho}V_\mu,$$

$$\{V_\mu, V_\nu\} = -S_{\mu\nu},$$

$$\{S_{\mu\nu}, Q\} = -\frac{1}{4}[\gamma_\mu, \gamma_\nu]Q, \quad \{V_\mu, Q\} = \frac{1}{2}\gamma_\mu Q \tag{7.1.6}$$

The Poincaré transformation law (7.1.1) for Q is seen to be a canonical one, with $S_{\mu\nu}$ as generators.

Let us turn to requirement (ii). The Q term in L is already Poincaré invariant--it is even SO(3,2) invariant. Clearly L' can depend only on $\dot{x}$ and Q, in a Lorentz invariant way. There are no Lorentz scalars we can form out of Q: $\tilde{Q}\beta Q$, $\tilde{Q}\beta\gamma_5 Q$, $\tilde{Q}\beta\gamma_5\gamma_\mu Q$ all vanish, and any scalars we might form with $S_{\mu\nu}$ and V_μ also vanish (see (7.1.3). We find that in addition to $\dot{x}^2$, the only scalar we can make up is $\dot{x}^\mu V_\mu$: all other possitilities do not yield anything independent. Thus L' has to be a function of $\dot{x}^2$ and $\dot{x}.V$.

The Q term in L in eq. (7.1.4) already respects chronometric invariance, being homogeneous of degree one in velocities. Extending this property to L', we find that the most general Lagrangian with spinor internal variable is

$$L = \frac{1}{2}\dot{\tilde{Q}}\beta Q + \sqrt{-\dot{x}^2}\quad f(\zeta),$$
$$\zeta = \dot{x}.V/\sqrt{-\dot{x}^2} \tag{7.1.7}$$

It is specified by one arbitrary function of one argument: this makes the spinorial model algebraically much simpler than the vectorial one.

§2. PHASE SPACE, CONSERVATION LAWS AND FORMS OF CONSTRAINT

Generators of Poincaré Group. To express the dynamics corresponding to the Lagrangian (7.1.7) in phase space form, we need to introduce a total momentum four-vector P_μ canonically conjugate to x_μ:

$$P_\mu = \frac{\partial L}{\partial \dot{x}^\mu} = f'V_\mu + \frac{1}{\sqrt{-\dot{x}^2}}(\zeta f' - f)\dot{x}_\mu \qquad (7.2.1)$$

The total phase space is twelve dimensional and the non-vanishing brackets are

$$\{x_\mu, P_\nu\} = g_{\mu\nu}, \qquad \{\Omega_a, \Omega_b\} = \beta_{ab} \qquad (7.2.2)$$

Poincaré invariance of L leads to the conservation of

$$M_{\mu\nu} = x_\mu P_\nu - x_\nu P_\mu + S_{\mu\nu}, \quad P_\mu, \qquad (7.2.3)$$

and these obey the PB relations for the Lie algebra of the Poincare group (see eq. (6.2.18). $S_{\mu\nu}$ here is formed from Q as in eq. (7.1.2). The Pauli-Lubanski vector W_μ is

$$W_\mu = \frac{1}{2}\varepsilon_{\mu\nu\rho\sigma}P^\nu S^{\rho\sigma} \quad . \qquad (7.2.4)$$

With the help of the identities (7.1.3) we are able to express its square in the form

$$W^2 = (P.V)^2 \qquad (7.2.5)$$

Constraints. In this model the only primary constraint we expect is the one due to chronometric invariance. It must involve Lorentz scalars formed from P_μ and Ω: these are immediately seen to be P^2 and P.V. From eq. (7.2.1), and the fact that V is lightlike, these two scalars are the following functions of ζ.

$$P^2 = (\zeta f')^2 - f^2, \; P.V = \zeta(\zeta f' - f) \quad . \qquad (7.2.6)$$

We define the generic case as obtaining when both P^2 and P.V are not constants; otherwise we have an exceptional case. The latter turn out to correspond to f of the form

$$f(\zeta) = c_1\zeta + c_2/\zeta, P^2 = -4c_1c_2, \quad P.V = -2c_2 \quad . \qquad (7.2.7)$$

In fact with the expressions (7.2.6), the constancy of P^2 implies that of P.V and vice versa. If f is generic, we eliminate ζ between P^2 and P.V and get a primary constraint Φ which we write in the form

$$\Phi = P^2 - \alpha(P.V) \approx 0 \qquad (7.2.8)$$

with α a nontrivial function of its argument. Thus the one function $f(\zeta)$ in L goes over into one function $\alpha(P.V)$ in Φ. With the result (7.2.5) we recognize the constraint (7.2.8) as a relation between the two Casimir invariants of the Poincaré algebra.

Classical limit of Majorana equation not included. We examine the possible forms of Φ a bit further, to see to what extent we can produce classical Lagrangian models corresponding to the Majorana and new Dirac equations. A classical limit of the Majorana equation would correspond to obtaining just the single constraint P.V = constant; but this is unattainable in our formalism based on a Lagrangian, since we will automatically get P^2 constant as well. We can at best get a projection onto a single mass-spin level of Majorana's equation, and in a sense this is what the new Dirac equation achieves. One can ask next if a generic f can lead to Φ of the form

$$\Phi = P^2 - a_o\, P.V - a_1\, (P.V)^2 = 0 \qquad (7.2.9)$$

but again this turns out impossible, given the forms (7.2.6) of P^2 and P.V. We can produce the relation (7.2.9) only by f being exceptional and P^2, P.V. separately being constants. Thus in particular trajectories with mass linear in spin are excluded.

§3. EQUATIONS OF MOTION AND SPACE-TIME TRAJECTORY

Hamiltonian. We have as Hamiltonian an arbitrary multiple of the primary constraint,

$$H = v\Phi\ , \qquad (7.3.1)$$

and get the following equations of motion for x, P and Q:

$$\begin{aligned} \dot{x}_\mu &\approx v\{x_\mu, \Phi\} \approx v(2P_\mu - \alpha'(P.V)V_\mu), \\ \dot{P}_\mu &\approx 0, \\ \dot{Q} &\approx v\{Q,\Phi\} \approx -\frac{1}{2} v\alpha'(P.V)\, P^\mu \gamma_\mu Q. \end{aligned} \qquad (7.3.2)$$

The equations for $S_{\mu\nu}$, V_μ can be found from the one for Ω or more directly by computing their PB's with H:

$$\dot{S}_{\mu\nu} = v\{S_{\mu\nu}, \Phi\} = -v\alpha'(P.V)(P_\mu V_\nu - P_\nu V_\mu),$$

$$\dot{V}_\mu = v\{V_\mu, \Phi\} = v\alpha'(P.V)\, S_{\mu\nu}P^\nu . \tag{7.3.3}$$

Dirac brackets identical to Poisson brackets. In this model there is only one constraint: $\Phi \approx 0$ and it is therefore first class. Hence, the constraint analysis is rather simple. The Dirac brackets are identical to the PB's (7.2.2), in particular $\{x_\mu, x_\nu\}^* = 0$. This contrast to the vector model of the previous chapter where besides the first class constraint corresponding to $\Phi \approx 0$, one has two second class constraints $P\cdot a \approx 0$, $P\cdot b \approx 0$ causing at this stage $\{x_\mu, x_\nu\}^*$ to be nonzero, see (6.3.18). Thus the present model avoids a serious factor ordering problem which one has to face when quantifying the vector model (or the model of Appendix A) in interaction with the electromagnetic field. [On quantizing one replaces $\{x_\mu, x_\nu\}^*$ by the corresponding commutator. But as $\{x_\mu, x_\nu\}^* \neq 0$ in the vector model, how is one to order the noncommuting components of the four vector x in $A_\mu(x)$ and $F_{\mu\nu}(x)$?]

Solution to equation of motion. Thus, the Dirac brackets are just the PB's (7.2.2) (later on this will be changed when we introduce a gauge constraint). In this way the equations of motion can be solved, and there appears one unknown function $\phi(s)$ due to the chronometric invariance. Recognizing that P^2 and P.V are constants of motion, we find that x, Q, S and V evolve in this way:

$$x_\mu(S) = x_\mu(0) + \left(\frac{2}{\alpha'(P.V)} - \frac{P.V}{P^2}\right)\frac{P_\mu}{\sqrt{-P^2}}\,\phi(s) - \frac{1}{P^2}\Big(S_{\mu\nu}(s) - S_{\mu\nu}(0)\Big)P^\nu ,$$

$$Q(s) = \left[\cos\left(\frac{\phi(s)}{2}\right) - \frac{P^\mu\gamma_\mu}{\sqrt{P^2}}\sin\left(\frac{\phi(s)}{2}\right)\right] Q(0) , \tag{7.3.4}$$

$$S_{\mu\nu}(s) = S_{\mu\nu}(0) - \frac{1}{\sqrt{-P^2}}(P_\mu V_\nu(0) - P_\nu V_\mu(0))\sin\phi(s) +$$

$$+ \frac{1}{P^2}(P_\mu S_{\nu\rho}(0) - P_\nu S_{\mu\rho}(0))P^\rho(1 - \cos\phi(s)).$$

$$V_\mu(s) = P_\mu\frac{P.V}{P^2} + (V_\mu(0) - P_\mu\frac{P.V}{P^2})\cos\phi(s) + \frac{1}{\sqrt{-P^2}}S_{\mu\nu}(0)\;P^\nu\sin\phi(s).$$

Here $\phi(s)$ is determined by v through the differential equation

$$\dot{\phi}(s) = v\alpha'(P.V)\ \sqrt{-P^2}\ , \quad \phi(0) = 0 \tag{7.3.5}$$

and P_μ and Q(0) must of course be chosen to obey the constraint $\Phi \approx 0$.

The motion in space-time is qualitatively similar to the vectorial models: x_μ traces on the average a straight line parallel to P_μ, but superimposed on that is an oscillating component orthogonal to P_μ. Quite naturally, Q "rotates" at half the speed at which $S_{\mu\nu}$, V_μ and (the oscillating part of x_μ) do.

Spacelike orbits can be prevented. Unlike the two-variable model A of chapter 6, the present spinor model does not guarantee that for any choice of f in the Lagrangian, spacelike P_μ will be excluded. However, one can characterize, directly at the Hamiltonian level, choices of the function $\alpha(P.V)$ that are physically reasonable, and see that such do exist. Let us write $z = P.V$. The spinor model does ensure that W_μ is spacelike, since $W^2 = z^2$. We can ensure that P_μ is timelike if $\alpha(z)$ is a negative definite function of z for all real z. (The fact that P_μ itself appears in the definition of z does not matter). Next let us suppose for simplicity that $\alpha(z)$ is an even function, so that positive and negative timelike P_μ occur symmetrically. We can now ask what further conditions must be imposed on $\alpha(z)$ to make sure that the space-time trajectory $x^\mu(s)$ always has a timelike tangent. (In the case of model A in chapter 6, this too was guaranteed by the structure of the equations: from eqs. (6.3.8a) with $v_2 = 0$, $\dot{x}$ is proportional to P which is timelike!). From the $\dot{x}_\mu$ equation in (7.3.2) this is the requirement

$$\dot{x}^2 \approx 4\ v^2(\alpha(z) - z\alpha'(z)) < 0 \quad ,$$

that is,

$$\left(\frac{\alpha(z)}{z}\right)' \ > \ 0 \tag{7.3.6}$$

It is clear that the twin conditions (i) $\alpha(z)$ negative, (ii) $\frac{\alpha(z)}{z}$ monotonously increasing with z cannot both be satisfied if $\frac{\alpha(z)}{z}$ is continuous at $z = 0$. For then $\frac{\alpha(z)}{z}$ would have to increase smoothly from positive values for $z < 0$ to negative values for $z > 0$! If we allow $\frac{\alpha(z)}{z}$ to be discontinuous at $z = 0$, then both conditions can be met in the two regions $z < 0$ and $z > 0$. A simple example is

$$\alpha(z) \ = - z/\sinh z \ , \tag{7.3.7}$$

and others can easily be imagined. This discussion is only intended to show that by careful choice of the trajectory function α we can get a model which is quite satisfactory at the classical level.

§4. QUANTIZATION OF THE FREE SPINOR MODEL

Gauge constraint. As a prelude to discussing the quantization problem, we carry the classical analysis one step further and introduce a gauge constraint to lift the chronometric invariance. Thus we are treating the spinor model differently from the way we treated the vectorial model in paragraph 6.5, and will give up manifest relativistic invariance. This will permit a clearer discussion of the electromagnetic interaction in the next subsection.

We adopt the gauge constraint

$$\chi = x^o - s \approx 0 \tag{7.4.1}$$

to serve as a conjugate to the existing primary constraint $\Phi \approx 0$. That is, χ and Φ form a second class pair. The maintenance of (7.4.1) fixes the hitherto arbitrary coefficient v in the Hamiltonian:

$$\frac{d\chi}{ds} = \frac{\partial\chi}{\partial s} + \{\chi,\Phi\}v \approx 0 \Rightarrow v = (2P^o - V^o\ \alpha'(P.V))^{-1}. \tag{7.4.2}$$

The general equation of motion for any dynamical variable f becomes

$$\frac{df}{ds} = \frac{\partial f}{\partial s} + (2P^o - V^o\alpha'(P.V))^{-1}\ \{f,\Phi\} \tag{7.4.3}$$

We can eliminate the two constraints $\chi = 0$, $\Phi = 0$ entirely by passing to the DB's

$$\{f,g\}^* = \{f,g\} + (2P^o - V^o\alpha'(P.V))^{-1}[\{f,x^o\}\{\Phi,g\} - \{f,\Phi\}\{x^o,g\}] \tag{7.4.4}$$

This is a bracket defined on a ten dimensional phase space for which $\vec{x}$, $\vec{P}$, Q may be taken as fundamental variables. The nonvanishing DB's among them are

$$\{x_j,\ P_k\}^* = \delta_{jk}\ , \qquad \{Q_a,\ Q_b\}^* = \beta_{ab} \tag{7.4.5}$$

Notice the crucial fact that the $\{x_j,\ x_k\}^*$ brackets vanish, in contrast to the vectorial model. The equation $\Phi = 0$, which is a strong one once we change from the use of PB's to DB's, can be supposed solved for P^o as a function of other variables:

$$\vec{P}^2 - (P^o)^2 - \alpha(\vec{P}.\vec{V} - P^oV^o) = 0 \Rightarrow$$

$$P^o = H(V^o, \vec{P}.\vec{V}, \vec{P}2). \qquad (7.4.6)$$

This H is the ordinary Hamiltonian in the sense that it generates the equation of motion (7.4.3) through the DB: for any dynamical variable expressed as a function of $\vec{x}$, $\vec{P}$, Ω and t (=x^o),

$$\frac{df}{dt} = \frac{\partial f}{\partial t} + \{f, H\}^* . \qquad (7.4.7)$$

Relativistic invariance is now expressed by the fact that the ten generators $M_{\mu\nu}$, P_μ yield a DB realization of the Poincaré algebra. It is natural to exhibit them in three-dimensional form as

$$J = \vec{x}_\wedge\vec{P} + \vec{S}, \qquad \vec{K} = \vec{x}\, H - t\,\vec{P} + \vec{S}' , \qquad (7.4.8)$$

$$\vec{P},\ H(V^o, \vec{P}.\vec{V}, \vec{P}^2)$$

$\vec{J}$ and $\vec{K}$ generate spatial rotations and pure Lorentz transformations respectively. $\vec{S}$ is the triplet of space-space components of $S_{\mu\nu}$, and $\vec{S}'$ is the triplet of time-space components.

Quantum Theory. To set up a corresponding relativistic quantum theory involves two steps. One first converts the classical DB's (7.4.5) into commutation relations for operators: thus one seeks hermitian operators $\vec{x}$, $\vec{P}$, Ω among which the only non-vanishing commutators are

$$[x_j, P_k] = i\hbar\,\delta_{jk}\ , \quad [\Omega_a, \Omega_b] = i\hbar\beta_{ab} \qquad (7.4.9)$$

These are easy to solve: one gets an irreducible representation in the Hilbert space H_0 of chapter 2. This means we have chosen $\vec{x}$ diagonal, and P acts as -ih $\vec{\nabla}$. For H_0 we may choose any representation, for instance the one with q_1 and q_2 diagonal. The Hilbert space H and its inner product,

$$||\,\psi\,|| = \int d^3x(\psi(\vec{x}),\psi(\vec{x}))_{H_0} \qquad (7.4.10)$$

are appropriate for the usual quantum mechanical interpretation. The second, and really nontrivial, step is the following: one must find a way of ordering the noncommuting factors V^0, $\vec{P}.\vec{V}$ in H, such that the expressions (7.4.8) close under commutation to yield a representation of the Poincaré algebra. The real difficulty lies in ensuring that H and

$$\vec{K} = \frac{1}{2}(\vec{x}\,H + H\,\vec{x}) + \vec{S}' \tag{7.4.11}$$

will obey

$$[K_j, H] = i\hbar\, P_j, \quad [K_j, K_k] = -i\hbar\,\epsilon_{jk\ell} J_\ell \quad . \tag{7.4.12}$$

If this problem can be solved, then we have a relativistic quantum system which shows the multiplicity-free spin spectrum $s = 0, \frac{1}{2}, 1, \ldots$, with mass a function of spin.

Dirac form vs. Foldy-Wouthuysen form. It is useful to compare this description with the one developed in chapter 3. The situation is analogous to the relationship between the Dirac and the Foldy-Wouthuysen-Thomas forms for the usual electron wave equation. Giving the Poincaré generators for the present system in the form (7.4.8) is like giving them for the Dirac equation in the form

$$\vec{P} = \vec{p} = -ih\vec{\nabla}\ , \quad H = \vec{\alpha}.\vec{p} + \beta m\ ,$$
$$\vec{J} = \vec{x}\wedge\vec{p} - \frac{i\hbar}{\mu}\vec{\alpha}\wedge\vec{\alpha}\ , \quad \vec{K} = \vec{x}\,H - i\hbar/2\vec{\alpha} - t\vec{p} \quad . \tag{7.4.13}$$

The Hamiltonian is not diagonal, but in this form electromagnetic interactions are easy to introduce. (See the next section.) On the other hand, the description of chapter 3 is similar to giving the generators for the Dirac equation as

$$\vec{P} = \vec{P},\ H = \beta\sqrt{\vec{p}^2 + m^2}\ ,$$
$$\vec{J} = \vec{x}\wedge\vec{p} + \frac{\hbar}{2}\vec{\sigma},\quad \vec{K} = \tfrac{1}{2}\{\vec{x}, H\} + \frac{\hbar}{2}\ \frac{\vec{p}\times\vec{\sigma}}{H + \beta m} \quad . \tag{7.4.14}$$

§5. RELATION TO THE MODELS OF CHAPTERS 3 AND 4

To establish the quantized free model of chapters 3 and 4 we proceed in a way which seems possible only for free particles. This way to proceed solves the factor ordering problem involved in writing the Hamiltonian corresponding to (7.4.6). First we go back to the manifestly covariant form of the classical theory, as discussed in §7.1, §7.2, §7.3. By quantizing this model in a straight forward way one obtains the model of chapter 4. How do we get the model of chapter 3? We assume that the function $\alpha(P.V)$ has been chosen such that classically only timelike P_μ occur. We shall assume that these P_μ

are restricted to the forward lightcone, but this is not essential for what follows. To every timelike P_μ there is a unique boost $(B(P))$ which boosts the four vector $(M,0,0,0)$ to P_μ , where $P^2 = -M^2$. This boost is defined to be the Lorentz-transformation in the plane spanned by P_μ and the time axis. Now, make a canonical transformation on the harmonic oscillator coordinates corresponding to this boost with the use of the generator $S_{o\nu}$ of (7.1.2). This transformation corresponds to *aligning* the oscillators with P and clearly is different for those P which are not on the same ray through the origin (the P's which have the same four velocity). This canonical transformation changes P.V into MV_o with $M = \sqrt{-P^2}$ and changes the generators of the Poincare group from (7.2.3). The generators P_μ are unaffected by the oscillator transformation, it is less obvious that the M_{ij} (i,j = 1,2,3) are unaffected also. The generators $M_{oj} = K_j$ are changed in that the part which operates on the oscillators is changed from S_{oi} to $\frac{\vec{P}\wedge\vec{S}}{P^0 + M}$. The generators P_μ satisfy $P^2 + M^2 = 0$ with now $M^2 = \alpha(MV_o)$ which can be solved as $M = h(V_o)$. Next, with this classical system, one proceeds as in §7.4 and introduces the constraint $\chi = x^o - S \approx 0$. Compared with §7.4 the simplification is that now one has

$$\vec{P}^2 - P^{o2} = \alpha(MV_o), \text{ i.e.}$$

$$M^2 = (MV_o), \text{ solving this formally}$$

$$\text{by } M = h(V_o), \text{ one has} \tag{7.5.1}$$

$$P^o = \sqrt{\vec{P}^2 + h^2(V_o)}$$

instead of (7.4.6). As P and V_o will commute there are no factor ordering problems in (7.5.1). In this way one finds again the generators (3.4.6) and the wave functions with innerproduct (3.4.2).

§6. ELECTROMAGNETIC INTERACTIONS

Classical Spinor Model. For the classical spinorial model, interaction with an external electromagnetic field, corresponding to a potential $A_\mu(x)$, is introduced by adding the usual term to the free Lagrangian (7.1.7). Thus the complete Lagrangian is

$$L = \frac{1}{2}\dot{Q}^{\sim}\beta Q + \sqrt{-\dot{x}^2}\, f(\zeta) + eA_\mu(x)\dot{x}^\mu . \tag{7.6.1}$$

Note that both gauge and chronometric invariances are maintained. The canonical momentum P_μ gets altered by the interaction:

$$P_\mu = \frac{\partial L}{\partial \dot{x}^\mu} = f'V_\mu + \frac{1}{\sqrt{-\dot{x}^2}}(\zeta f' - f)\dot{x}_\mu + eA_\mu(x) ; \tag{7.6.2}$$

writing $\pi_\mu = P_\mu - eA_\mu(x)$, we must clearly find the primary constraint Φ now by eliminating ζ between

$$\begin{aligned} \pi^2 &= (\zeta f')^2 - f^2 , \\ \pi . V &= \zeta(\zeta f' - f) . \end{aligned} \tag{7.6.3}$$

It follows that Φ is formed in just the same way as in the free case, using the same trajectory function α:

$$\Phi = (P - eA(x))^2 - \alpha((P - eA(x)).V) \approx 0 \tag{7.6.4}$$

In principle, the classical analysis goes through with no essential problems. We only note that on adopting again the gauge constraint $\chi = x^o - s = 0$, we get DB's that depend on the interaction:

$$\{f,g\}^* = \{f,g\} + (2\Pi^o - V^o\alpha'(\Pi.V))^{-1}[\{f,x^o\}\{\Phi,g\} - \{f,\Phi\}\{x^o,g\}]. \tag{7.6.5}$$

However, the DB's among $\vec{x},\vec{P},Q$ are unaffected and remain as before, eq. (7.4.5). And now the Hamiltonian that generates the equations of motion via the DB is

$$H' = H(V^o, \vec{\Pi}.\vec{V}, \vec{\Pi}^2) + eA^o(\vec{x},t) . \tag{7.6.6}$$

In other words, the Hamiltonian description of the free spinor model developed in the previous subsection is the right one in which the electromagnetic interaction is properly described by the minimal replacement principle.

On quantization, the $\vec{x}$ become commuting operators, so provided the factor ordering problems can be solved we have no problems of interpretation. This is in sharp contrast to the vectorial model of chapter 6 and the Hanson-Regge spherical top model of Appendix A.

The Hamiltonian (7.6.6) is most interesting in that, due to the presence of the internal vector $\vec{V}$, it can generate nontrivial magnetic moments for all the particles lying on the mass-spin trajectory. These moments must ultimately depend on the trajectory relation itself.

In the next section we present now a classical calculation to show how this comes about.

§7. CALCULATION OF THE MAGNETIC MOMENTS OF THE STATES

The field free motion (7.3.4) gives a helix in space time with P_μ giving the direction of the screw and the x^μ follows its thread. The charge is located at x^μ, i.e. on the thread. Measured in units of proper time the frequency of this helical motion is MS^{-1}, where M is the mass $M = \sqrt{-P^2}$, and S the average spin in units $\hbar$. The radius of the helical motion is $M^{-1}S$. The total angular momentum is constant ; it consists of the orbital part, $L_{\mu\nu}$, and of the spin $S_{\mu\nu}$, given in (7.2.3); neither of these is separately conserved. One may say that the internal spinor variables keep the particle away from its average position and force the helical motion. Such a motion generates a magnetic moment, the calculation of which is the subject of this section. Although this calculation is completely classical, there is a close analogy with what happens for the Dirac equation. To exhibit this, we first present a calculation of the g-factor for the electron in a somewhat unusual way, the role of the helical motion being played by the Zitterbewegung.

Rederivation of the electron g factor. Dirac's original derivation of the electron g-factor involves squaring the Hamiltonian and cannot be adapted to our model, therefore we give here an alternative approach.

We use the Heisenberg picture and set $\hbar = c = 1$. In the presence of a homogeneous external magnetic field $\vec{B}$, the operator equations of motion are generated by the Hamiltonian

$$H = H_0 + H_{int}, \quad H_0 = \vec{\alpha}.\vec{p} + \beta m, \; H_{int} = -e\vec{\alpha}.\vec{A}(\vec{x}),$$

$$A(x) = \frac{1}{2}\vec{B}\wedge\vec{x}. \tag{7.6.7}$$

The operators, $\vec{\alpha}$, $\vec{x}$,... appearing here are of course time dependent. In its present form H_{int} does not yet involve terms like $\vec{L}.\vec{B}$ and $\vec{S}.\vec{B}$, where $\vec{L}$ is the orbital angular momentum and the spin $\vec{S}$ is

$$\vec{S} = \frac{1}{2}\vec{\alpha} = -\frac{i}{4}\vec{\alpha}\wedge\vec{\alpha} \;. \tag{7.6.8}$$

We shall now show that if one averages over the oscillatory motion (Zitterbewegung) of the electron, and takes the low momentum limit, H_{int} is actually of the physically expected form.

Working to lowest order in eB, we may assume that in

$$H_{int} = \frac{e}{2}\,\vec{B}.\vec{\alpha}(t)\wedge\vec{x}(t) \tag{7.6.9}$$

the time dependences in $\vec{\alpha}$ and $\vec{x}$ correspond to free motion due to H_o. Following Schrödinger and Dirac [DIR 3], this motion is explicitly known: $\vec{p}$ and H_o are constant, while $\vec{\alpha}$ and $\vec{x}$ at any two times t, t + Δt are connected by

$$\vec{\alpha}(t+\Delta t) = \frac{\vec{P}}{H_0} + (\vec{\alpha}(t)-\frac{\vec{P}}{H_0})\,e^{-2iH_0\,t} = \frac{\vec{P}}{H_0} + e^{2iH_0\,t}(\vec{\alpha}(t)-\frac{\vec{P}}{H_0})\ ,$$

$$\vec{x}(t+\Delta t) = \vec{x}(t) + \frac{\vec{P}}{H_0}\,t + \frac{e^{2iH_0\,t}-1}{2iH_0}(\vec{\alpha}(t) - \frac{\vec{P}}{H_0}$$

$$= \vec{x}(t) + \frac{\vec{P}}{H_0}\,\Delta t + (\vec{\alpha}(t) - \frac{\vec{P}}{H_0})\ \frac{1-e^{-2iH_0\Delta t}}{2iH_0}\ .$$

Superposed on the uniform motion is the high frequency Zitterbewegung. This makes H_{int} in (7.6.9) a rapidly fluctuating function of time. We define a short-term average of $\vec{\alpha}(t)\wedge\vec{x}(t)$ as the average of $\alpha(t+\Delta t)\wedge x(t+\Delta t)$ with respect to Δt over many cycles of the Zitterbewegung centered about t: in this process, terms $e^{\pm 2iH_0\Delta t}$ and $te^{\pm 2iH_0\Delta t}$ are discarded. The result will of course depend on t, but the high frequency terms will have been removed. Denoting this average by a bar, we have after a simple calculation:

$$H_{int} = \frac{e}{2}\,\vec{B}.\overline{\vec{\alpha}(t)\ \vec{x}(t)} = -\frac{e}{2}\,\vec{B}\ (\frac{1}{H_0}\,(\vec{L}(t)+2\vec{S}(t)) + \frac{i}{2H_0}\,\vec{\alpha}(t)\wedge\vec{p}). \tag{7.6.10}$$

For a slowly moving electron, the matrix elements of $\vec{\alpha}(t)$ between positive energy states are proportional to $\vec{p}$, as is clear from

$$\vec{\alpha}H_0 + H_0\vec{\alpha} = 2\vec{p}. \tag{7.6.11}$$

In that limit, we may drop the term $\vec{a}\wedge\vec{p}$ as being quadratic in $\vec{p}$, and also replace H_0 by m. This yields

$$H_{int} = -\frac{e}{2m}\,\vec{B}.(\vec{L}+2\vec{S}), \tag{7.6.12}$$

which shows that the g-factor for the Dirac electron is two, up to the approximations made.

Magnetic moment of Regge sequence. The calculation for our spinor model although classical and not quantum mechanical is closely analogous to the calculation of the g-factor for the electron as given above.

For a homogeneous (weak) external magnetic field B the vector potential is

$$\vec{A} = \frac{1}{2}\,\vec{B}\wedge\vec{x}, \qquad A_o = 0 \quad . \tag{7.6.12}$$

We expand the total Hamiltonian H' of (7.6.6) about its free value H, retain only terms linear in $\vec{B}$, and ultimately pass to the extreme nonrelativistic limit. Let us write u,v,w, for the arguments V^0, $\vec{P}.\vec{V}$, $\vec{P}^2$ of H. From eq. (7.4.6), whose solution gives us H(u,v,w), we find:

$$\begin{pmatrix} \dfrac{\partial H}{\partial u} \\[2ex] \dfrac{\partial H}{\partial v} \\[2ex] \dfrac{\partial H}{\partial w} \end{pmatrix} = \frac{1}{2H-u\alpha'(z)} \begin{pmatrix} H\alpha'(z) \\[2ex] -\alpha'(z) \\[2ex] 1 \end{pmatrix} \quad , \quad z = P.V \tag{7.6.13}$$

Thus to first order in $\vec{B}$, H' is given by:

$$H' = H(u,v,w) + H_{int} \quad ,$$

$$H_{int} = \frac{e}{2c}\,\vec{V}.\vec{B}\wedge\vec{x}\,\frac{H}{v} - \frac{e}{c}\,\vec{P}.\vec{B}\wedge\vec{x}\,\frac{\partial H}{\partial w}$$

$$= -\frac{e}{c}\,\frac{1}{2H-u\alpha'(z)}\,\left(\vec{L} - \frac{\alpha'(z)}{2}\,\vec{x}\wedge\vec{V}\right).\vec{B} \quad . \tag{7.6.14}$$

Here $\vec{L}$ is the orbital angular momentum. For the coefficient of $\vec{B}$ here we will use the solutions (7.3.4) to the free equations of motion. Then H, w, and z are all constants of the motion. Now both $x_\mu(s)$ and $V_\mu(s)$ have periodic "time" dependences in their free motion, in addition to the uniform motion in $x_\mu(s)$.

We next follow the previous method for handling the unfamiliar term $\vec{x}\wedge\vec{V}$, and replace it by its short-term average. As we are only working to lowest order in e, we may again compute this average on the basis of free motion, which is described by (7.3.4). Thus we form the product $x(s+\Delta s)\wedge V(s+\Delta s)$ from (7.4.7), and drop all oscillatory terms like $\sin\Delta s$, $\cos\Delta s$, $\sin\Delta s\cos\Delta s$, $\Delta s\sin\Delta s$, $\Delta s\cos\Delta s$, and obtain a result which of course depends on s:

$$\vec{x}(s)\wedge\vec{V}(s) = -\frac{z}{M^2}\,\vec{x}(s)x\wedge\vec{P} - \frac{2z}{M^4}\,\vec{P}\wedge\overline{S(s).\vec{P}}$$

$$-\frac{\vec{V}(s)}{M^2}\wedge\overrightarrow{S(s).P} \qquad (7.6.16)$$

The last term here can be simplified using

$$V_\lambda S_{\mu\nu} + V_\mu S_{\nu\lambda} + V_\nu S_{\lambda\mu} = 0 \quad .$$

We then drop terms quadratic in $\vec{P}$, as well as terms linear in S_{oj} because in the quantized version in the Majorana representation of $S_{\mu\nu}$, the S_{oj} have vanishing $\Delta j=o$ matrix elements.
Then we find

$$\vec{x}(s)\wedge\vec{V}(s) = -\frac{z}{M^2}\,(\vec{L}(s) - \vec{S}(s)), \qquad (7.6.17)$$

and

$$H_{int} = -\frac{e}{2H_o + w\alpha'}[1 + \frac{z\alpha'}{2M^2})\,\vec{L}(s) - \frac{z\alpha'}{2M^2}\,\vec{S}(s)]\,.\vec{B}\,. \qquad (7.6.18)$$

Considering the limit $P\to 0$, one has

$$(2H_o + w\alpha')^{-1}(1 + \frac{z\alpha'}{2M^2}) \longrightarrow (2H_o + V_o\alpha')^{-1}(1+\alpha'\frac{H_oV_o}{2H^2} = \frac{1}{2H_o} \quad ,$$

$$-(2H_o + w\alpha')^{-1}\,\frac{z\alpha'}{2M^2} \longrightarrow -(2H_o + V_o\alpha')^{-1}\,\frac{\alpha' V_o}{2H_o} = \frac{V_o}{2H_o^2}\,\frac{\partial H_o}{\partial V_o}. \qquad (7.6.19)$$

For H_{int} one finds the result:

$$H_{int} = -\frac{e}{2H_o}\,(\vec{L}(s) + \frac{V_o}{H_o}\,\frac{\partial H_o}{\partial V_o}\,\vec{S}(s)).\vec{B}\,, \qquad (7.6.20)$$

i.e., the gyromagnetic factor is given by [VAN 8].

$$g = \left.\frac{\partial \ln H_o}{\partial \ln V_o}\right|_{u=v=o} \quad . \qquad (7.6.21)$$

Thus our result for the present model is that the g-factor is determined by the function f in (7.1.7). Let the Regge relation implied by this same f be $M=M_o\,\beta(S)$, where S is again average spin/h. Then our final result is

$$g(S) = \frac{d\ln\beta(S)}{d\ln S}$$

Hence if we assume $\beta(S) = S^n$, then $g(S) = n$; i.e. for the usual assumption $n = 1/2$, $g(S) = 1/2$ for all states of the Regge trajectory. This is in striking contrast to the well-known result $g(S) = \frac{1}{S}$ obtained from finite component wave equations describing one spin value at a time [HAG 1].

§8. RELATIVISTIC SU(6) MODEL, STRING MODEL

The Lagrangian (7.6.1) contains in Q and in $\zeta = \dfrac{\dot{x}^\mu V_\mu}{\sqrt{-\dot{x}^2}}$, a pair of degenerate harmonic oscillator variables. A fully relativistic SU(6) model with a minimal electromagnetic interaction is easily obtained by replacing the single pair of degenerate harmonic oscillators by a triplet of such pairs. One then has a six degenerate harmonic oscillator; this is the model discussed in VAN 3. The mass is given via $P^2 = \alpha(P.(V^{(1)} + V^{(2)} + V^{(3)}))$, where the single V_μ based on one pair is now replaced by the sum of three V_μ's, one each for each of the triplets of pairs of oscillator variables. The "spin", (cf. also Chapter 3 §7) is given via a similar sum over the triplet of pairs: $S_{\mu\nu} = S^{(1)}_{\mu\nu} + S^{(2)}_{\mu\nu} + S^{(3)}_{\mu\nu}$. The rest mass of the free particle is a function only of the total number of quanta $n = n^{(1)} + n^{(2)} + n^{(3)}$. The spin involves an angular term of the angular momenta in the three degenerate modes. Thus, for n = 0 one has a spin zero singlet; the n=1 state is an SU(3) triplet of spin 1/2 states; n = 2 has a singlet with spin 1 and an antitriplet with spin 0; n = 3 is the familiar SU(6) 56-plet. The SU(6) symmetry is broken easily by replacing P.V by a sum of $P.V^{(i)}$ using different constants in front of the three $V^{(i)}$. (That this model does not contradict O'Raifeartaigh's theorem was discussed in Chapter 4 §3.) Let us here give explicitly the interaction with an external electromagnetic field. The Lagrangian is

$$L = \sum_{i=1}^{3} \frac{1}{2}\dot{Q}^{(i)T} \beta Q^{(i)} + \sqrt{-\dot{x}^2}\, f(\zeta) + eA_\mu \dot{x}^\mu , \tag{7.8.1}$$

where

$$\zeta = \frac{1}{\sqrt{-\dot{x}^2}}\, \dot{x}^\mu (c^{(1)} V^{(1)}_\mu + c^{(2)} V^{(2)}_\mu + c^{(3)} V^{(3)}_\mu) \tag{7.8.2}$$

where the symmetry is broken via the (in principle unequal) $c^{(i)}$. Some more remarks on the classical version of this model are contained in §4 of Chapter 8.

String Model. The model given by the Lagrangian (7.6.1) can also be generalized to produce the spectrum of the string model [BDV 1]. To do this one replaces the single pair of harmonic oscillator variables with a denumerably infinite set of pairs of such variables. The Lagrangian (7.6.1) then reads, labelling these sets of pairs by N:

$$L = \sum_{N=1}^{\infty} \frac{1}{2} \dot{\Omega}^{(N)^T} \beta\Omega^{(N)} + \sqrt{-\dot{x}^2}\, f(\zeta) + eA_\mu \dot{x}_\mu \,, \tag{7.8.3}$$

where now

$$\zeta = \frac{1}{\sqrt{-\dot{x}^2}} \dot{x}^\mu \Big(\sum_{N=1}^{\infty} N V_\mu^{(N)} \Big). \tag{7.8.4}$$

CHAPTER EIGHT

FURTHER ANALYSIS OF THE CLASSICAL MOTION OF THE SPINOR MODEL

§1. DYNAMICS IN TERMS OF PRIMARY VARIABLES

Here we derive the dynamical equations in terms of the variables $x^\mu(s)$, $\Omega(s)$, $V^\mu(s)$, $S^{\mu\nu}(s)$, used in chapter 7, in the general case of interaction with an external electromagnetic field. The motion is in terms of s, as yet an arbitrary parameter for the world line.

The Hamiltonian conjugate to s is as given in chapter 7:

$$H = v(s)\ \{(P-eA)^2 - \alpha[(P-eA)\cdot V]\}\ . \qquad (8.1.1)$$

Introduce

$$\Pi_\mu = P_\mu - eA_\mu\ . \qquad (8.1.2)$$

where P_μ is conjugate to x^ν: $\{x^\nu, P_\mu\} = \delta^\nu{}_\mu$, i.e. ,

$$\{\Pi_\mu, \Pi_\nu\} = -eF_{\mu\nu}. \qquad (8.1.3)$$

Using the Poisson bracket rules given in chapter 7 for V_μ, V_ν, $S_{\mu\nu}$, one has with (8.1.1) as Hamiltonian:

$$\dot{\Pi}_\mu = v[-2eF_{\mu\nu}\Pi^\nu + \alpha' F_{\mu\nu}V^\nu]; \qquad (8.1.4)$$

$$\dot{x}_\mu = v[2\Pi_\mu - \alpha' V_\mu], \text{ i.e.} \qquad (8.1.5)$$

$$\dot{\Pi}_\mu = -veF_{\mu\nu}\dot{x}^\nu\ ; \qquad (8.1.6)$$

$$\dot{S}_{\mu\nu} = v\alpha'\ [\Pi_\mu V_\nu - \Pi_\nu V_\mu] \qquad (8.1.7)$$

$$\dot{V}^\mu = -v\alpha' e S_{\mu\nu}\Pi^\nu \qquad (8.1.8)$$

$$(\Pi^2)^{\dot{}} = 2v\alpha' eF_{\alpha\beta}\Pi^\alpha V^\beta \qquad (8.1.9)$$

$$(\Pi.\Gamma)^{\dot{}} = -2veF_{\alpha\beta}\Pi^\alpha V^\beta \qquad (8.1.10)$$

§2. FREE PARTICLE KINEMATICS, INTRODUCTION OF SECONDARY COORDINATES AND SPIN

There are ten conserved quantities corresponding to the ten generators of the Poincaré group. These quantities may be derived via Noether's theorem from the Lagrangian, as done in chapter 7 or by analogy straight-forwardly from the Hamiltonian (8.1.1). By inspection the conserved quantities are for $A_\mu = 0$.

$$P_\mu = \frac{1}{2v}\dot{x}_\mu + \frac{\alpha'}{2} V_\mu \tag{8.2.1}$$

$$M_{\mu\nu} = x_\mu P_\nu - x_\nu P_\mu + S_{\mu\nu} \tag{8.2.2}$$

where in (8.2.1),(8.1.5) was used.

For the free particle one might be tempted to identify $S_{\mu\nu}$ as the spin. This however is wrong for at least two reasons. The first reason is that $S_{\mu\nu}$ is lightlike, one has (7.1.6) $S_{\mu\nu}S^{\mu\nu} = 0$ and thus it can not be perpendicular to the timelike P_μ. The second reason is that $S_{\mu\nu}$ is not separately conserved, one has, with (8.1.7),

$$\dot{S}_{\mu\nu} = -(x_\mu P_\nu - x_\nu P_\mu)^{\cdot} = v\alpha' \cdot (P_\mu V_\nu - P_\nu V_\mu). \tag{8.2.3}$$

The problem here is that x_μ *is not the center of mass* of our spinning particle.

Center of Mass, General Procedure (cf. Pryce [PRY 1]. To find this one proceeds as follows, using the conserved P_μ and $M_{\mu\nu}$ of (8.2.1), (8.2.2). First go to a rest frame of the particle; in that frame $P_\mu = (\sqrt{-P^2}, 0,0,0)$. Second, in the Euclidean space of the rest frame (which is orthogonal to its time axis), choose the origin so that $M_{oi} = 0$. The possibility of such a choice in the rest frame, $P_\mu = (\sqrt{-P^2},0,0,0)$, follows from the fact that in that system a shift δx_i of origin gives a change in M_{oi} of

$$\delta M_{oi} = -\sqrt{P^2}\ \delta x_i \text{ (rest frame)}. \tag{8.2.4}$$

Thus one finds a center of mass frame of coordinates which is unique apart from a freedom of rotation and time translation. In this center of mass system, one has again $P_\mu = (\sqrt{-P^2},0,0,0)$, and also the spatial *center of mass position* of the particle is at the origin (0,0,0) for all values of the world line parameter s; indeed, we define center of mass in this way. Thus it follows that the *world line of the center of mass coincides with the time axis in a center of mass frame.* Note that for this result all that one needs is the ten conservation laws of P_μ and $M_{\mu\nu}$. Actually we only need those for P_μ and M_{oi}, the M_{ij} will show that the *spin* is conserved also.

Let us apply this to our case; first we find a rest frame and consider the event on the world line for which $x^o = 0$. Next, we shift the Euclidean origin over y_i changing M_{oi} to $M_{oi}{}^y = 0$, with

$$0 = M_{oi}{}^y = -(x_i - y_i)\sqrt{-P^2} + S_{oi}, \text{ i.e.}$$

$$y_i = x_i - S_{oi}/\sqrt{-P^2} \quad . \tag{8.2.5}$$

Hence, the *center of mass* of the particle in a Euclidean space perpendicular to P is *not at* x_i *but at* y_i given by (8.2.5).

What is the spin in the rest frame? Clearly this is what remains of $M_{\mu\nu}$ in a rest frame, for our model this is just S_{ij} with $i,j = 1,2,3$. In other words the *spin* $\Sigma_{\mu\nu}$ *in a rest frame is given by*

$$\Sigma_{ij} = S_{ij} \quad i,j = 1,2,3 \text{ (rest frame)} \tag{8.2.6}$$

$$\Sigma_{io} = 0 \ . \tag{8.2.7}$$

That this spin $\Sigma_{\mu\nu}$ is conserved in a rest frame follows from the conservation laws of P_μ and $M_{\mu\nu}$.

Covariant Notation, Secondary Coordinates. What we did in a rest frame generalizes in a general frame to the introduction of center of mass coordinates y_μ and of spin $\Sigma_{\mu\nu}$, where

$$y_\mu = x_\mu - S_{\mu\nu}P^\nu/(-P^2). \tag{8.2.8}$$

Note, however, that unlike the x_μ which satisfy $\{x_\mu, y_\nu\} = 0$, the y_μ have nonvanishing Poisson Brackets. Introducing explicitly the shift vector

$$d_\mu = S_{\mu\nu}P^\nu/(-P^2), \tag{8.2.9}$$

$$\Sigma_{\mu\nu} = S_{\mu\nu} + d_\mu P_\nu - d_\nu P_\mu \ . \tag{8.2.10}$$

The total angular momentum can be written

$$M_{\mu\nu} = (y_\mu P_\nu - y_\nu P_\mu) + \Sigma_{\mu\nu}, \tag{8.2.11}$$

where the two terms on the right hand side are now separately conserved:

$$\dot{M}_{\mu\nu} = 0 \quad , \quad \dot{\Sigma}_{\mu\nu} = 0. \tag{8.2.12}$$

Also, by inspection or by transforming from a rest frame:

$$P^{\mu}\Sigma_{\mu\nu} = 0. \tag{8.2.13}$$

Alternatively, one could define $\Sigma_{\mu\nu}$ as a translation-invariant antisymmetric tensor formed linearly from $M_{\mu\nu}$ and orthogonal to P_μ,

$$\Sigma_{\mu\nu} = M_{\mu\nu} + \frac{1}{P^2}(P_\mu M_{\mu\lambda} - P_\nu M_{\mu\lambda})P^\lambda .$$

Eqs. (8.2.10,11) then follow easily.

Pauli-Lubanski Vector. This is a useful alternative to $\Sigma_{\mu\nu}$.

$$W = \frac{1}{2}\varepsilon^{\mu\nu\lambda\tau}P_\nu\Sigma_{\lambda\tau} = \frac{1}{2}\varepsilon^{\mu\nu\lambda\tau}P_\nu S_{\lambda\tau} \tag{8.2.14}$$

clearly

$\dot{W}^{\mu} = 0$, the inverse of (8.2.14) is

$$\Sigma^{\mu\nu} = \frac{1}{-P^2}\varepsilon^{\mu\nu\alpha\beta} P_\alpha W_\beta . \tag{8.2.15}$$

Although they are equivalent it is useful to have both Σ and W available.

§3. FREE PARTICLE MOTION

The equations of motion for no external field are:

$$\dot{P}_\mu = 0, \quad \dot{x}_\mu = v[2P_\mu - \alpha' V_\mu]$$

$$\dot{S}_{\mu\nu} = v\alpha'[P_\mu V_\nu - P_\nu V_\mu], \quad \dot{V}_\mu = -v\alpha' S_{\mu\nu}P^\nu . \tag{8.3.1}$$

To analyze this motion in detail we introduce the center of mass coordinate y_μ, the shift d_μ of (8.2.8) and (8.2.9):

$$x_\mu = y_\mu + d_\mu \tag{8.3.2}$$

$$d_\mu = \frac{S_{\mu\nu}P^\nu}{(-P^2)} \tag{8.2.9}$$

The equations of motion for these secondary coordinates follow from (8.3.1); there is an important check as it follows, as discussed in §2, from the conservation laws that $\dot{y}_\mu$ will be parallel to P_μ. One finds

$$\dot{d}_\mu = \frac{\dot{S}_{\mu\nu}P^\nu}{(-P^2)} = v\alpha'(V_\mu + \frac{P.V}{(-P^2)}P_\mu) . \tag{8.3.3}$$

Because of (7.1.6) the vector V_μ is a null vector and it is useful to introduce the components of V_μ in the direction of $P_\mu \equiv h_\mu$ and perpendicular to $P_\mu \equiv f_\mu$:

$$h_\mu = -\frac{P.V}{(-P^2)} P_\mu \text{ , and} \tag{8.3.4}$$

$$f_\mu = V_\mu - h_\mu \tag{8.3.5}$$

where $h^2 + f^2 = 0$. With these variables we can write (8.3.3) as

$$\dot{d}_\mu = v\alpha' f_\mu \text{ ,} \tag{8.3.6}$$

with $\dot{V}_\mu$ from (8.3.1), one has

$$\dot{h}_\mu = 0 \text{ ,} \tag{8.3.7}$$

$$\dot{f}_\mu = \dot{V}_\mu = -\,v\alpha' d_\mu(-P^2) \text{ .} \tag{8.3.8}$$

Pictorial Analysis of the Motion. This is done in figure 8.1. First we check that y_μ describes a straight world line parallel to P_μ, as is necessary because of the arguments of §2. With (8.3.1), (8.3.2), (8.3.6) we find

$$\dot{y}_\mu = \dot{x}_\mu - \dot{d}_\mu = vP_\mu[2 - \alpha'\frac{(-P.V)}{(-P^2)}] \text{ .} \tag{8.3.9}$$

Here the v represents the arbitrariness still present in the choice of the parameter s along the world line $x_\mu(s)$.

Equations (8.3.6) and (8.3.8) make it clear that the two vectors d_μ and $f_\mu/\sqrt{-P^2}$, which are both orthogonal to P_μ, and to each other (because of (7.1.3)), run around each other on a circle in a plane perpendicular to P_μ:

$$\dot{d}_\mu = v\alpha' \sqrt{-P^2}\ (f_\mu/\sqrt{-P^2}) \tag{8.3.10}$$

$$(f_\mu/\sqrt{-P^2})^{\cdot} = -v\alpha' \sqrt{-P^2}\ d_\mu .$$

Thus, $x_\mu = y_\mu + d_\mu$, *the position of the charge describes* as s proceeds *a spiral around the straight world line of the center of mass* y_μ. The radius of the spiral is given by $\sqrt{d^2}$, the size of which we shall discuss shortly; this is illustrated in figure 8.2.

The Spin $\Sigma_{\mu\nu}$ and also $S_{\mu\nu}$ and the Pauli Lubanski vector W_μ can be expressed in terms of the secondary variables:

$$\Sigma_{\mu\nu} = -(d\wedge f)_{\mu\nu} \frac{\sqrt{-P^2}}{\sqrt{\Sigma^2}} \text{ ,} \tag{8.3.11}$$

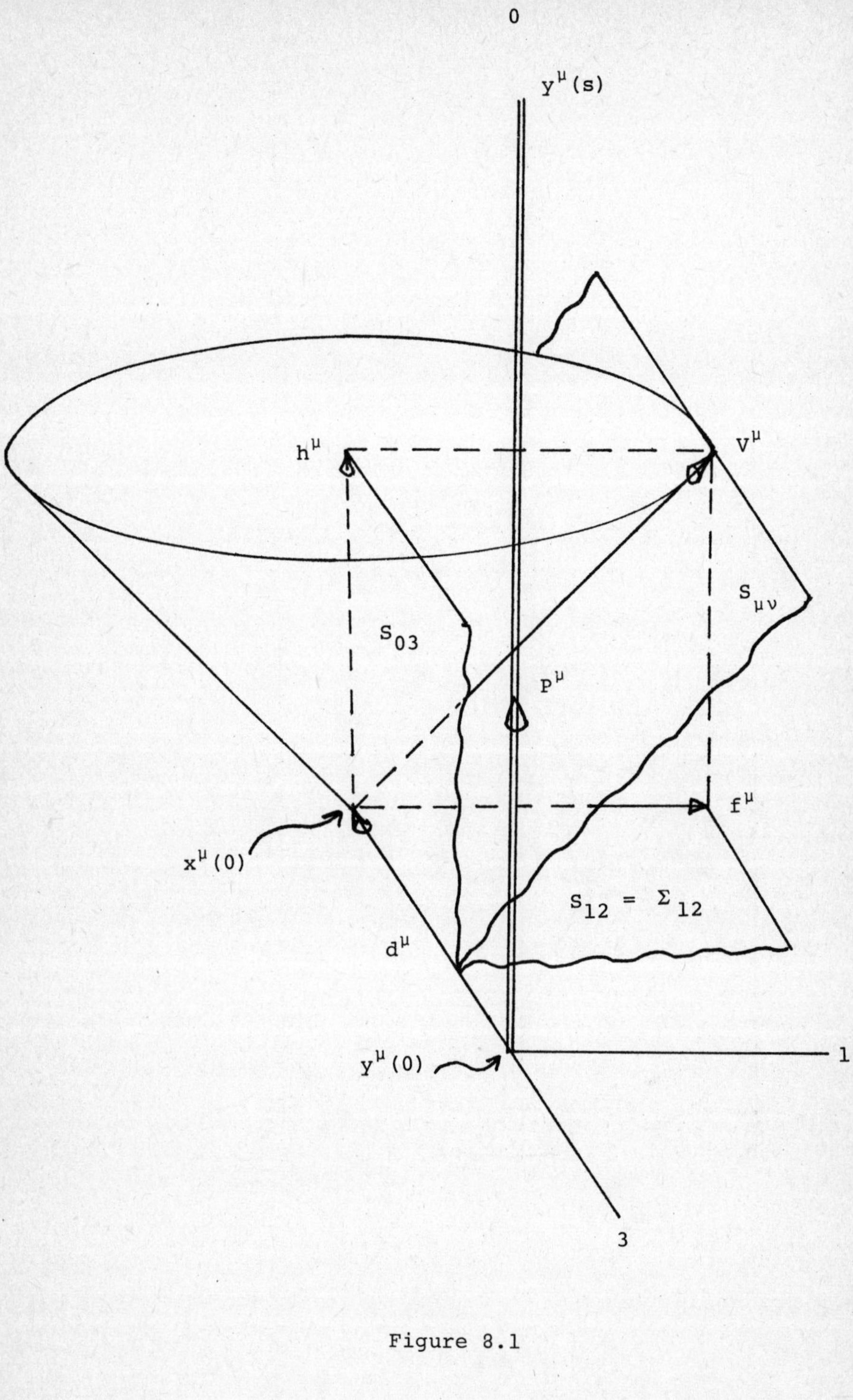

Figure 8.1

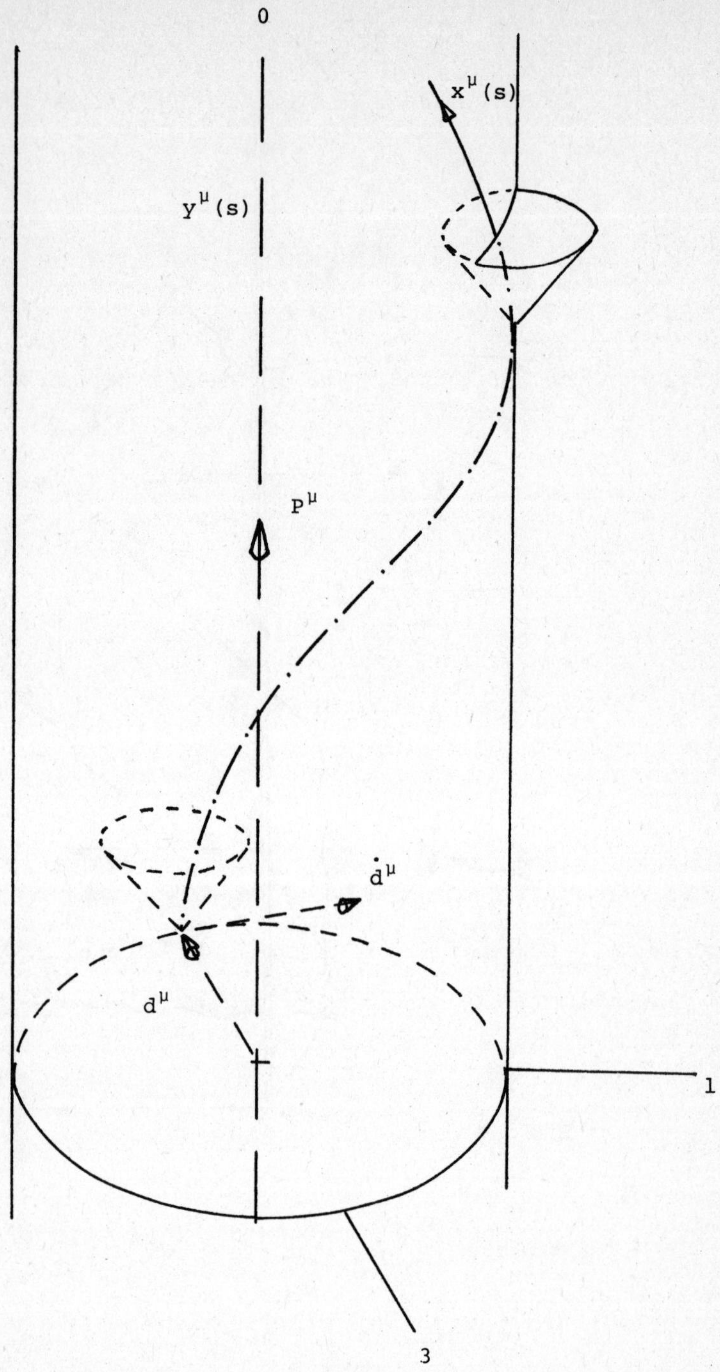

Figure 8.2

where $\Sigma^2 = \frac{1}{2}\,\Sigma_{\mu\nu}\Sigma^{\mu\nu} = \dfrac{(P.V)^2}{(-P^2)}$; (8.3.12)

in which we used $S_{\lambda\mu}S^{\lambda}{}_{\nu} = V_\mu V_\nu$ from (7.1.3) ;

$$S_{\mu\nu} = -\ (d_\wedge V)_{\mu\nu}\ \frac{\sqrt{-P^2}}{\sqrt{\Sigma^2}}\ ; \tag{8.3.13}$$

$$W_\mu = \frac{1}{2}\varepsilon_{\mu\nu\lambda\tau}\ P^\nu d^\lambda V^\tau\ \frac{\sqrt{-P^2}}{\sqrt{\Sigma^2}}\ , \tag{8.3.14}$$

all this is illustrated in figure 8.1. Recalling (7.1.3) we have

$S^{\mu\nu}S_{\mu\nu} = V^\mu V_\mu = 0$,i.e. $S_{\mu\nu}$ is a tangent plane to the lightcone and V_μ is a null vector. Also, $\varepsilon_{\mu\nu\lambda\tau}S^{\nu\lambda}V^\tau = 0$, i.e. V_μ is in the plane $S_{\mu\nu}$, as well as perpendicular to it: $V^\mu S_{\mu\nu} = 0$.

For the nonzero lengths, we have

$$\frac{1}{2}\,\Sigma^{\mu\nu}\Sigma_{\mu\nu} = \Sigma^2,\ P^\mu P_\mu = P^2,\ (P.V)^2 = \Sigma^2(-P^2),$$

$$d^2 = \frac{\Sigma^2}{(-P^2)}\ ,\ -h^2 = f^2 = \Sigma^2,$$

$$W^2 = \Sigma^2(-P^2)\ , \text{ and finally the} \tag{8.3.15}$$

Regge relation of chapter 7:

$$P^2 = \alpha(P.V) = \alpha(\sqrt{-P^2\Sigma^2}) \tag{8.3.16}$$

Choosing a parameter s *along the world line.* The parameter s, used to label $x_\mu(s)$, is free so far. But in order to answer the question: what is the *frequency of the spiral motion?* We must now fix it. As $\dot{y}_\mu$ is parallel to P_μ and thus to a time axis of an inertial frame at all times, it is reasonable to set our standard of time by the demand that $\dot{y}_\mu\dot{y}^\mu = -1$, where $\dot{y}_\mu = \frac{d}{ds}\,y_\mu(s) = \frac{d}{ds}(y_\mu(s) - d_\mu(s))$. From (8.3.9) this determines v(s) via

$$-1 = \dot{y}^2 = -\ {}^2(-P^2)\,[2 - (\frac{(-P.V)}{(-P^2)}\,\alpha']^2, \text{ or}$$

$$v = -\ \frac{\sqrt{-P^2}}{[P.V\alpha' - 2P^2]}\ . \tag{8.3.17}$$

Using (8.3.15) and denoting $\sqrt{-P^2}$ by M one has

$$v = \frac{M}{M\Sigma\alpha'(M\Sigma) - 2\alpha(M\Sigma)}\ . \tag{8.3.18}$$

The frequency of the circular motion in space is now determined from the motion of d and $f/\sqrt{-P^2}$, given by (8.3.10) as

$$\omega = v\alpha' M = \frac{\alpha' M^2}{M\Sigma\alpha'(M\Sigma) - 2\alpha(M\Sigma)} \tag{8.3.19}$$

Spin is orbital motion which is due to the spinorial variables keeping the particle from its average position.

With (8.3.11) we have an equation which can be written as

$$\Sigma_{\mu\nu} = -(d_{\wedge} f/\sqrt{-P^2})_{\mu\nu} \frac{(-P^2)}{\sqrt{\Sigma^2}} = (\dot{d}_{\wedge} d)_{\mu\nu} \frac{M^2}{\Sigma} \tag{8.3.20}$$

which substantiates our claims that the spin is due to an orbital motion.

4. SPIN FOR PARTICLE IN AN EXTERNAL ELECTROMAGNETIC FIELD, GYROMAGNETIC RATIO

When passing from a free particle to a particle in interaction with an electromagnetic field, we replace P_μ by $\Pi_\mu = P_\mu - eA_\mu$ in the definition of the secondary variables such as $\Sigma_{\mu\nu}$ which we used in §2 and in §3, for instance

$$d_\mu(s) = \frac{S_{\mu\nu}\Pi^\nu}{(-\Pi^2)} \quad . \tag{8.2.9a}$$

One quickly notices $d_\mu\Pi^\mu = 0$, $\Sigma_{\mu\nu}\Pi^\nu = 0$, $f_\mu\Pi^\mu = 0$, as before, but now Π^2 and Σ^2 are no longer constants, c.f. (8.1.9), (8.1.10).

What interests us most here are the motions of $\Pi_\mu(s)$ and of $\Sigma_{\mu\nu}(s)$, for these quantities. The instantaneous equations are already given by (8.1.4)-(8.1.10). However, the electromagnetic field couples to the charge which is located at the event x_μ which spirals rapidly around the world line of the center of mass y_μ. Thus one gets radiation, but even neglecting that, a rather complicated motion. The motion becomes relatively simple only if one averages over the rapid spiraling motion of the charge. This is only allowed if the field is sufficiently weak, also in its derivatives. This averaging can be done in two ways. The first way is to obtain the average of the interaction term in the Lagrangian and then use a result [VAN 1] which gives the motion of Π_μ and of $\Sigma_{\mu\nu}$ directly from the ten conservation laws of linear and angular momentum for particle plus field. The second way to proceed, which we carry out also, is to calculate $\dot{\Pi}^\mu$ and $\dot{\Sigma}^{\mu\nu}$ using (8.1.4) - (8.1.10) and then average over the rapidly spiralling motion.

Averaging the interaction over the internal motion. We have (7.6.1)

$$S_I = e \int d^4x \int ds\delta(\bar{x} - x(s)]\ \dot{x}^\mu(s) A_\mu(x),$$

in this one substitutes

$$x^\mu(s) = y^\mu(s) + d^\mu(s), \text{ with}$$

$$d^\mu(s) = \frac{S^{\mu\nu}\Pi_\nu}{(-\Pi^2)}, \text{ this gives}$$

$$S_I = e \int ds\ \{[\dot{y}^\mu(s) + \dot{d}^\mu(s)]\ A_\mu[y(s)] +$$

$$+ [\dot{y}^\mu(s) + \dot{d}^\mu(s)]\ A_{\mu,\alpha}[y(s) d^\alpha(s) + \ldots \qquad (8.4.1)$$

where we have written three dots for the higher derivatives of the field A^μ. Neglecting these higher derivatives and assuming that eA^μ and its derivatives are small enough to neglect the perturbation due to eA^μ on x^μ and d^μ in (8.4.1) (second order in e), we average over the built in free rotational motion of $d_\mu(s)$ and obtain

$$\bar{S}_I = e \int ds\ \{\dot{y}^\mu(s)\ A_\mu[y(s)] + \overline{\dot{d}^\mu d^\alpha} A_{\mu,\alpha}[y(s)] + \ldots\ .$$

The results of §3, visualized in figures 1 and 2, gives

$$\overline{\dot{d}^\mu d^\alpha} = \tfrac{1}{2}\overline{(\dot{d}^\mu d^\alpha - d^\alpha \dot{d}^\mu)} = v\alpha'\Sigma\Sigma_{\mu\alpha}/M\ . \qquad (8.4.2)$$

Denoting the factor in front of $\Sigma_{\mu\alpha}$ by 2g we have, neglecting higher derivatives of the field and using the antisymmetry of Σ

$$\bar{S}_I = d \int ds\ \{\dot{y}^\mu(s)\ A_\mu[y(s)] + g\ \Sigma^{\mu\alpha}\ (A_{\mu,\alpha} - A_{\alpha,\mu})\ . \qquad (8.4.3)$$

This interaction term identifies a particle with a charge e at position $y^\mu(s)$ and with a magnetic moment $g\Sigma^{\mu\nu}$, i.e. g is the gyromagnetic ratio. Fixing v(s) via $\dot{y}^2 = -1$, with (8.3.18), one has for g with (8.4.2)

$$g = \frac{1}{4M}\ \frac{M\Sigma\alpha'(M\Sigma)}{M\Sigma\alpha'(M\Sigma) + 2\alpha(M\Sigma)}\ . \qquad (8.4.4)$$

Note that only on average does the particle appear to have just charge and a magnetic moment (provided one neglects effects of order e^2 and higher derivatives of the field). (A charge at distance $\vec{d}$ from $\vec{y}$

actually contains all moments, of course.) At this point we could use the results of [VAN 1] to give us the average motion of Π^μ and of $\Sigma^{\mu\nu}$ directly in terms of the source term (8.4.3). Let us do instead the more complicated direct calculation.

Let us first calculate $\dot{y}_\mu = \dot{x}_\mu + \dot{d}_\mu$, $\dot{x}_\mu$ being given by (8.1.5), and $\dot{d}_\mu$ by

$$\dot{d}_\mu = \frac{2v\alpha' eF_{\alpha\beta}\Pi^\alpha V^\beta}{(-\Pi^2)} d_\mu + v\alpha' f_\mu + ev[\Pi_\mu d_\alpha - \Pi_\alpha d_\mu]F^{\alpha\beta}[-2\Pi_\beta + \alpha' V_\beta]. \tag{8.4.5}$$

This gives

$$\dot{y}_\mu = \Pi_\mu(2v - \alpha' \frac{-\Pi.V}{-\Pi^2}) + d_\mu(\frac{2v\alpha' eF_{\alpha\beta}\Pi^\alpha V^\beta}{-\Pi^2})$$
$$+ ev[\Pi_\mu d_\alpha - \Pi_\alpha d_\mu]F^{\alpha\beta}[-2\Pi_\beta + \alpha' V_\beta]. \tag{8.4.6}$$

Note that in the presence of an external field $\dot{y}_\mu$ is no longer parallel to Π_μ, there are two additional terms each of which is perpendicular to Π and proportional to e.

Turning now to the equations of motion which concern us most: $\dot{\Pi}_\mu$ and $\dot{\Sigma}_{\mu\nu} = \dot{S}_{\mu\nu} + \dot{d}_\mu\Pi_\nu - \dot{d}_\nu\Pi_\mu + d_\mu\dot{\Pi}_\nu + d_\nu\dot{\Pi}_\mu$, we have from (8.1.6)

$$\dot{\Pi}_\mu = -veF_{\mu\nu}(y+d)(\dot{y}^\nu + \dot{d}^\nu) = -v[eF_{\mu\nu}(y)(\dot{Y}^\nu + d^\nu) +$$
$$+ eF_{\mu\nu,\beta}(y)(\dot{y}^\nu + \dot{d}^\nu)(y^\beta + d^\beta) + \ldots] . \tag{8.4.7}$$

Next we average, neglecting the higher derivatives of A_μ, and neglecting in the motion of x and d the influence of the fields (order e^2), one finds

$$\overline{\dot{\Pi}}_\mu = -veF_{\mu\nu}(y)\dot{y}^\nu - evF_{\mu\nu,}\overline{\tfrac{1}{2}(d^\beta\dot{d}^\alpha - d^\alpha\dot{d}^\beta)},$$

again using §3 one has

$$\overline{\dot{\Pi}}_\mu = -veF_{\mu\nu}(y)\dot{y}^\nu + \frac{ev}{2}(\frac{v\alpha'\Pi.V}{\alpha})F_{\mu\nu,\beta}(y)\Sigma^{\nu\beta} .$$

Using the antisymmetry of $\Sigma^{\nu\beta}$ and Maxwell's equation:

$F_{\mu\nu,\beta} + F_{\beta\mu,\nu} + F_{\nu\beta,\mu} = 0$, one may rewrite the last equation as

$$\overline{\dot{\Pi}}_\mu = -veF_{\mu\nu}(y)\dot{y}^\nu - \frac{ev}{4\sqrt{-\Pi^2}} gF_{\mu\nu,\beta}(y)\Sigma^{\nu\beta} . \tag{8.4.8}$$

Here one recognizes the gyromagnetic factor from the typical Stern-Gerlach term, again

$$g = \frac{\sqrt{-\Pi^2}\, v\alpha' \Pi . v}{\alpha} .$$

The equation for W_μ, which is equivalent to $\Sigma_{\mu\nu}$, is somewhat easier to handle, therefore we do it first. We have

$\dot{W}_\mu = \frac{1}{2} \varepsilon_{\mu\nu\lambda\tau} \dot{\Pi}^\nu S^{\lambda\tau}$, (the term with $\dot{S}^{\lambda\tau}$ drops out because of (8.1.7)). Averaging and neglecting higher derivatives and terms of order e^2 we obtain

$$\bar{\dot{W}}_\mu = \frac{eg}{\sqrt{-\Pi^2}} F_{\mu\nu} W^\nu + \frac{e}{2\sqrt{-\Pi^2}} (g-2) \dot{y}^\alpha F_{\alpha\beta} W^\beta \dot{y}_\mu$$

$$\frac{-eg}{-\Pi^2} W^\alpha \sigma_\alpha \tilde{F}_{\beta\gamma} W^\beta \dot{y}^\gamma \dot{y}_\mu , \tag{8.4.9}$$

where $\tilde{F}_{\beta\gamma}$ is the dual tensor to $F_{\mu\nu}$.

Similarly one finds

$$\dot{\Sigma}_{\mu\nu} = \dot{S}_{\mu\nu} + \dot{d}_\mu \Pi_\nu - \dot{d}_\nu \Pi_\mu + d_\mu \dot{\Pi}_\nu - d_\nu \dot{\Pi}_\mu =$$

$$= [d_\mu \Pi_\nu - d_\nu \Pi_\mu] [\frac{2v\sigma' e F_{\alpha\beta} \Pi^\alpha V^\beta}{-\Pi^2} - e F_{\alpha\beta} \Pi^\alpha \dot{x}^\beta]$$

$$- ev[d_\mu F_{\nu\alpha} \dot{x}^\alpha - d_\nu F_{\mu\alpha} \dot{x}^\nu] . \tag{8.4.10}$$

After using the by now familiar procedure of averaging and neglecting one gets

$$\bar{\dot{\Sigma}}_{\mu\nu} = \frac{eg}{2\sqrt{-\Pi^2}} [F_{\mu\alpha} \Sigma^\alpha{}_\nu - F_{\nu\alpha} \Sigma^\alpha{}_\mu] + \frac{e}{2\sqrt{-\Pi^2}} [\dot{y}_\mu \Sigma_{\nu\alpha} - \dot{y}_\nu \Sigma_{\mu\nu}] [(g-2) \dot{y}_\gamma F^{\gamma\alpha}$$

$$- \frac{g}{2\sqrt{-\Pi^2}} \sigma^\alpha F^{\gamma\sigma} \Sigma_{\gamma\sigma}] . \tag{8.4.11}$$

§5. ENERGY-MOMENTUM TENSOR

As we have a Lagrangian one might go ahead and blindly follow the standard procedure to obtain an energy momentum tensor. (WEY 1) In this way one gets incorrect results. The reason for this is that the standard procedure assumes that the variables involved are all local in the sense that they can be identified with a specific event in space-time. Our Lagrangian (7.1.7) involves such a local variable in $x^\mu(s)$,

However, it also contains a global variable Ω which is translation invariant and not attached to any point in particular. As $x^{\mu}(s)$ is the only local variable in the Lagrangian, one might attach Ω at the event $x_{\mu}(s)$. Then, however, the energy momentum tensor becomes localized at the point $x_{\mu}(s)$ and is of the form

$$T^{\mu\nu}(x) = \int ds\,\delta(x - x(s))\,a^{\mu\nu}(s) \; . \qquad (8.5.1)$$

It is clear, for the free particle, that (8.5.1) is not satisfactory, cf. §2, and figures 1 and 2. This is also clear from the formulas one finds that $T^{\mu\nu}(x)$ given by (8.5.1) is neither conserved nor symmetric. Also one does not have $P^{\mu} = \int T^{\mu o} d\vec{x}$, nor $M^{\mu\nu} = \int (x^{\mu}P^{\nu 0} - x^{\nu}P^{\mu 0})\,d\vec{x}$, where P^{μ} and $M^{\mu\nu}$ are given by the Lagrangian cf. (7.2.1), (7.2.8).

One way to proceed makes some sense. Consider the average $\overline{T}^{\mu\nu}$ of $T^{\mu\nu}$ of (8.5.1) over the rapid spiraling motion of $\delta(x - x(s))$. Then indeed one has

$$\delta_{\mu}\overline{T^{\mu\nu}} = 0, \quad P^{\mu} = \int \overline{T^{\mu o}}\; d\vec{x} \; ,$$

$$M^{\mu\nu} = \int (x^{\mu}\overline{T}^{\nu o} - x^{\nu}\overline{T}^{\mu o})\,d\vec{x} \; .$$

One might then stipulate that the gravitational field interact with $\overline{T}_{\mu\nu}$ instead of with $T_{\mu\nu}$. There are some suggestions which support this kind of approach [SAK 1],[MIS 1] and which are based on the idea that gravity is not a fundamental field, but should be compared to the macroscopic theory of elastic bodies. Like the elasticity theory then gravity should not be taken seriously at the microscopic level. Of course there remain the question whether the size of the oscillatory behavior of our particles is below the scale over which gravity might be expected to average.

For weak gravitational fields there is no problem, one may apply the averaging procedures used for the electro magnetic field and obtain equations analogous to those which Hojman [HOJ 1] found for the Hanson-Regge model.

§6. CLASSICAL SU(6) MODEL WITH THREE PAIRS OF OSCILLATORS

The Lagrangian of this model was discussed in chapter 7. The model has three pairs of oscillators, the pairs being denoted by (1), (2), (3). The Lagrangian is a straightforward generalization of (7.1.7), including the electromagnetic interaction it is

$$L = \sum_{i=1}^{3} \frac{1}{2}\dot{\Omega}^{(i)}\beta\Omega^{(i)} + \sqrt{\dot{x}^2}\, f\left(\frac{\dot{x}\cdot(V^{(1)}+V^{(2)}+V^{(3)})}{\sqrt{\dot{x}^2}}\right) + e\dot{x}^{\mu}A_{\mu}. \quad (8.6.1)$$

The analysis is entirely similar to that of §2 and §3. Instead of d_μ one has now three four vectors $d_\mu^{(i)}$ each of which is perpendicular to Π_μ:

$$d_\mu^{(i)} = \frac{S_{\mu\nu}^{(i)}\Pi^\nu}{(-\Pi^2)} \qquad i = 1,2,3 \qquad (8.6.2)$$

For a free particle one obtains from the ten conserved quantities P_μ and $M_{\mu\nu} = x_\mu P_\nu - x_\nu P_\mu + \sum_{i=1}^{3} S_{\mu\nu}^{(i)}$, a straight world line parallel to P_μ for the center of mass coordinate y_μ. One now has

$$x_\mu(s) = y_\mu(s) + \sum_{i=1}^{3} d_\mu^{(i)}(s). \qquad (8.6.3)$$

For $\dot{d}_\mu^{(i)}(s)$ one again finds a quantity $v\alpha' f_\mu^{(i)}(s)$ which is perpendicular to $d_\mu^{(i)}(s)$ and to P_μ. The results are illustrated in figure 8.3 where we have drawn the three dimensional space of the rest frame. The center of mass is at $\vec{y}$, the position of the charge is at $\vec{x}$ which differs from $\vec{y}$ by the vector sum of the three $\vec{d}^{(i)}$. The spin is proportional to

$$\sum_{i=1}^{(3)} \dot{\vec{d}}^{(i)} \wedge \vec{d}^{(i)} \quad .$$

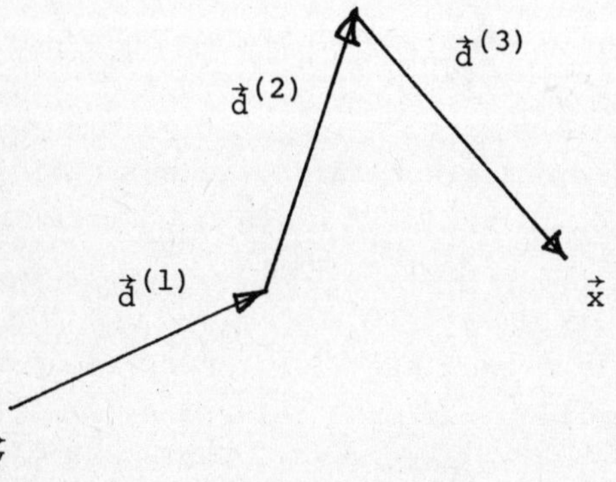

Figure 8.3

APPENDIX A:

THE RELATIVISTIC SPHERICAL TOP

The purpose of this appendix is to give a concise introduction to the classic work of Hanson and Regge [HAN 1], [HAN 2] on the relativistic spherical top. This will enable the reader to easily appreciate the similarities as well as important differences between our models, on the one hand, and the relativistic top on the other. In addition, the Hanson-Regge model is a very instructive application of Dirac's generalized Hamiltonian methods recounted in Chapter 5.

As in the body of these notes, we use the metric with $g_{00} = -1$: Hanson and Regge (H-R) use $g_{00} = 1$. Apart from this, H-R use a convention for the structure and action of the Poincaré group rather different from ours. We shall explain and use their formalism in this appendix, making it easier for the reader to refer to their paper. Our development of the details will, however, differ slightly from the original.

Elements of the Poincaré group are denoted (M,a), with $M = (M^{\mu}{}_{\nu})$ being an $SO(3,1)$ matrix, and a^{μ} a real four-vector. The composition law is taken as

$$(M,a)(M',a') = (MM',a' + M'^{-1}a) \quad ,$$

$$(MM')^{\mu}{}_{\nu} = M^{\mu}{}_{\rho}M'^{\rho}{}_{\nu} \quad , \quad (M^{-1})^{\mu}{}_{\nu} = M_{\nu}{}^{\mu} \quad . \tag{A.1}$$

Two inertial frames θ and θ' will be related by the group element (M,a), and this is indicated by $\theta' = \theta(M,a)$, if the coordinates x^{μ}, x'^{μ} assigned in θ, θ' respectively to a space-time event are connected thus:

$$\theta' = \theta(M,a): \quad x'^{\mu} = (M^{-1}x)^{\mu} + a^{\mu} = M_{\nu}{}^{\mu}x^{\nu} + a^{\mu} \quad . \tag{A.2}$$

Equations (A.1, A.2) are consistent in the sense that

$$(\theta(M,a))(M',a') = \theta((M,a)(M',a')) \quad . \tag{A.3}$$

The basic Lagrangian coordinates for the relativistic spherical top are a space-time position vector $x^{\mu}(s)$ and an $SO(3,1)$ matrix $\Lambda^{\mu}{}_{\nu}(s)$, both functions of an evolution parameter s. The configuration space is ten dimensional, and may be identified with the Poincaré group manifold. Physically, the top traces a world line in space-time, carrying a variable Lorentz frame with it. The action of the Poincaré group is given by:

$$\theta' = \theta(M,a): \quad x \to x' = M^{-1}x + a, \quad \Lambda \to \Lambda' = \Lambda M. \tag{A.4}$$

Thus x transforms as usual, while each row of Λ is a translation-invariant four-vector. In addition, an internal Lorentz group $SO(3,1)^{int.}$ can be defined to act on x, Λ in this way:

$$M \epsilon\ SO(3,1)^{int.}: \quad x \to x' = x, \quad \Lambda \to \Lambda' = M\Lambda\ . \tag{A.5}$$

Now x is invariant, and each column of Λ is a four-vector. The Poincaré group and $SO(3,1)^{int.}$ have commuting actions.

The true Lagrangian velocity variables are $\dot{x}^\mu \equiv u^\mu$, and the s-derivatives of the six independent parameters specifying $\Lambda(s)$, once some parametrization is chosen for SO(3,1). Instead of explicitly introducing a parametrization, one can use the relativistic angular velocity corresponding to $\Lambda(s)$: this is a second rank antisymmetric tensor $\sigma^{\mu\nu}$ defined by

$$\sigma^{\mu\nu} = -\sigma^{\nu\mu} = (\Lambda^{-1}\dot{\Lambda})^{\mu\nu} = \Lambda_\rho{}^\mu \dot{\Lambda}^{\rho\nu}\ . \tag{A.6}$$

The use of velocity type variables u, σ rather than true Lagrangian velocities leads to some changes in the canonical formalism, similar to what was seen in the vectorial model in Chapter 6. These are explained below. Under the two groups of transformations, u and σ behave as follows:

$$\theta' = \theta(M,a): \quad u'^\mu = (M^{-1}u)^\mu = M_\nu{}^\mu u^\nu\ ,$$

$$\sigma'^{\mu\nu} = (M^{-1}\sigma M)^{\mu\nu} = M_\lambda{}^\mu\, \sigma^{\lambda\rho} M_\rho{}^\nu\ ; \quad \text{(a)} \tag{A.7}$$

$$M \epsilon\ SO(3,1)^{int.}: \quad u' = u, \quad \sigma' = \sigma\ . \quad \text{(b)}$$

The Lagrangian L for the relativistic spherical top is subject to the following four independent requirements: (i) it must be manifestly Poincaré invariant; (ii) it must also be manifestly $SO(3,1)^{int.}$ invariant; (iii) it must possess chronometric invariance; (iv) on introducing canonical conjugates to x and Λ, to pass to a phase space description, L must lead to a set of primary constraints written as a vector equation $V_\mu \approx 0$: the form of V_μ is taken up later.

Let us first consider conditions (i) and (ii). Translation invariance means L may depend only on Λ, u and σ. Of these, Λ transforms under $SO(3,1)^{\text{int.}}$ as given by eq. (A.5), while both u and σ are $SO(3)^{\text{int.}}$ scalars (i.e., each component u^μ and $\sigma^{\mu\nu}$ is an $SO(3,1)^{\text{int.}}$ scalar). There are no nontrivial $SO(3,1)^{\text{int.}}$ scalars that one can form from Λ alone, so translation invariance and condition (ii) limit L to being a function of u and σ alone. (In other words, Λ may appear in only via its appearance in σ). Under the Poincaré group, u and σ are a four vector and an antisymmetric tensor respectively. Therefore the available Poincaré and $SO(3,1)^{\text{int.}}$ scalars are the four combinations:

$$a_1 = u^\mu u_\mu \quad , \quad a_2 = \sigma^{\mu\nu}\sigma_{\mu\nu} \quad ,$$

$$a_3 = u_\mu \sigma^{\mu\nu}\sigma_{\nu\lambda}u^\lambda \quad , \quad a_4 = \text{Det}\,\sigma = \frac{1}{16}(\sigma^{\mu\nu}\sigma^*_{\mu\nu})^2 . \tag{A.8}$$

Here we have defined σ^*, the dual to σ, as

$$\sigma^*_{\mu\nu} = \frac{1}{2}\,\varepsilon_{\mu\nu\lambda\rho}\,\sigma^{\lambda\rho} \quad , \qquad \varepsilon_{0123} = 1. \tag{A.9}$$

Conditions (i) and (ii) will be met by taking L to be any function of a_1, a_2, a_3, a_4. Imposing condition (iii) of chronometric invariance, we see that L must be of the form

$$L = \sqrt{-a_1}\; f\,(\xi,\eta,\zeta),$$

$$\xi = \frac{a_2}{a_1} \quad , \quad \eta = \frac{a_3}{{a_1}^2} \quad , \quad \zeta = \frac{a_4}{{a_1}^2} \quad , \tag{A.10}$$

with f any function of three arguments. Requirement (iv) on L will then be a restriction on f: it is better to examine this after introducing the phase space formalism.

The phase space Γ is of dimension twenty, and the (generalized) conjugates to x and Λ are defined by

$$P_\mu = \frac{\partial L}{\partial u^\mu} \quad , \quad S_{\mu\nu} = \frac{\partial L}{\partial \sigma^{\mu\nu}} \tag{A.11}$$

The partial derivative with respect to $\sigma^{\mu\nu}$ is interpreted in this sense: if one makes a variation $\delta\sigma_{\mu\nu}$ in $\sigma_{\mu\nu}$, maintaining antisymmetry in μ and ν, and this causes a change δL in L, one identifies $S_{\mu\nu}$ by writing

$$\delta L = \frac{1}{2}\, S_{\mu\nu}\,\delta\sigma^{\mu\nu} \quad , \tag{A.12}$$

it being understood that $S_{\mu\nu} = -S_{\nu\mu}$. To get the basic PB's on Γ, as well as the Euler Lagrange equations, we compute δL for an arbitrary infinitesimal variation in x(s) and Λ(s). Any variation of Λ(s) must take the form

$$\delta\Lambda^{\mu}{}_{\nu} = \Lambda^{\mu}{}_{\rho}\delta\theta^{\rho}{}_{\nu} \quad , \quad \delta\theta_{\mu\nu} = -\,\delta\theta_{\nu\mu}, \tag{A.13}$$

there being six independent quantities $\delta\theta$, since Λ must remain an SO(3,1) matrix. With this expression for the change in Λ one readily finds:

$$\delta\sigma_{\mu\nu} = \delta\dot{\theta}_{\mu\nu} + \sigma_{\mu}{}^{\rho}\delta\theta_{\rho\nu} - \delta\theta_{\mu}{}^{\rho}\sigma_{\rho\nu} \tag{A.14}$$

With these ingredients one has:

$$\begin{aligned}\delta L &= P_{\mu}\delta u^{\mu} + \frac{1}{2}\, S_{\mu\nu}\ \delta\sigma^{\mu\nu} \\ &= -\dot{P}_{\mu}\delta x^{\mu} - \frac{1}{2}(\dot{S}_{\mu\nu} - S_{\mu}{}^{\rho}\sigma_{\rho\nu} + \sigma_{\mu}{}^{\rho}S_{\rho\nu})\delta\theta^{\mu\nu} \\ &\quad + \frac{d}{ds}\,(P_{\mu}\delta x^{\mu} + \frac{1}{2}S_{\mu\nu}\delta\theta^{\mu\nu}).\end{aligned} \tag{A.15}$$

We may conclude from Hamilton's Principle that the Euler-Lagrange equations of motion are

$$\begin{aligned}&\dot{P}_{\mu} = 0 \quad , \\ &\dot{S}_{\mu\nu} = S_{\mu}{}^{\rho}\sigma_{\rho\nu} - \sigma_{\mu}{}^{\rho}S_{\rho\nu}, \quad \text{i.e,}\ \dot{S} = S\sigma - \sigma S.\end{aligned} \tag{A.16}$$

The total differential term in (A.15) fixes the PB's between x and Λ on the one hand, and P_{μ}, $S_{\mu\nu}$ on the other, since for arbitrary numerical δx, $\delta\theta$ we must have:

$$\begin{aligned}\delta x_{\mu} &= \{x_{\mu}, P_{\nu}\delta x^{\nu} + \tfrac{1}{2}S_{\nu\rho}\delta\theta^{\nu\rho}\} \quad , \\ \delta\Lambda^{\mu}{}_{\nu} &= \{\Lambda^{\mu}{}_{\nu}, P_{\rho}\delta x^{\rho} + \tfrac{1}{2}S_{\lambda\rho}\delta\theta^{\lambda\rho}\} \quad .\end{aligned} \tag{A.17}$$

To the consequences of these equations we may add the PB's among the $S_{\mu\nu}$ themselves, obtained by arguments similar to those of Section 7. In this way we find that the complete list of nonvanishing PB's among x, Λ, P and S is as follows:

$$\{x_\mu, P_\nu\} = g_{\mu\nu} \quad ; \quad \{\Lambda^\mu{}_\nu, S_{\alpha\beta}\} = \Lambda^\mu{}_\alpha\, g_{\nu\beta} - \Lambda^\mu{}_\beta g_{\nu\alpha} \; ;$$

$$\{S_{\mu\nu}, S_{\alpha\beta}\} = g_{\mu\alpha}\, S_{\nu\beta} - g_{\nu\alpha}\, S_{\mu\beta} + g_{\mu\beta}\, S_{\alpha\nu} - g_{\nu\beta}\, S_{\alpha\mu} \; . \qquad (A.18)$$

The Poincaré and $SO(3,1)^{int.}$ transformation laws for P_μ and $S_{\mu\nu}$ are as follows: when x, Λ change according to eqs. (A.4, A.5), then

$$\theta' = \theta(M,a): \quad P'_\mu = (M^{-1}P)_\mu = M^\nu{}_\mu P_\nu ,$$

$$S'_{\mu\nu} = (M^{-1}SM)_{\mu\nu} = M^\rho{}_\mu S_{\rho\lambda} M^\lambda{}_\nu \quad ; \quad (a)$$

$$M \,\varepsilon SO(3,1)^{int.}: \quad P'_\mu = P_\mu \; , \quad S'_{\mu\nu} = S_{\mu\nu} \qquad (b) \qquad (A.19)$$

It is useful to compare the tensor $S_{\mu\nu}$ in the H-R model with the corresponding tensors in the vectorial and spinorial models. In the vector case, eq. (6.2.15) ensures that there are only three algebraically independent components among $S_{\mu\nu}$; this, as we saw, must be so since the internal unit vector a_μ has just three algebraically independent components. In the spinor case, the six $S_{\mu\nu}$ are all built up from just four independent variables $\xi_1, \xi_2, \Pi_1, \Pi_2$ (see eq. (2.2.17), and the two identities in eq. (7.1.3) that involve $S_{\mu\nu}$ alone reflect this fact. In contrast, in the present case, there are no apriori kinematical restrictions on $S_{\mu\nu}$ at all, and all six components must be viewed as intrinsically independent quantities. Any constraints that may arise on $S_{\mu\nu}$ must be dynamical, i.e., they must be consequences of the singularity of L.

The invariance properties of L lead to corresponding conserved generators. From Poincaré invariance we find that

$$J_{\mu\nu} = x_\mu\, P_\nu - x_\nu\, P_\mu + S_{\mu\nu} \; , \quad P_\mu \qquad (A.20)$$

are constants of motion. Via their PB's they reproduce the Lie relations of the Poincare group: these are the same as eqs. (6.2.7) with the replacement $M_{\mu\nu} \to J_{\mu\nu}$. For the Pauli-Lubanski vector we have:

$$W_\mu = \frac{1}{2}\, \varepsilon_{\mu\nu\rho\lambda}\, P^\nu J^{\rho\lambda} = S^*_{\mu\nu}\, P^\nu$$

$$W^2 = \overline{V}^2 - \frac{1}{2}\, P^2 S^2 \; . \qquad (A.21)$$

Here S^* is dual to S, $S^2 = S_{\mu\nu}S^{\mu\nu}$, and the important vector $\overline{V}_\mu$ is defined as

$$\overline{V}_\mu = S_{\mu\nu} P^\nu . \qquad (A.22)$$

Thus V and W arise on "applying" S and S^* respectively to P. From the $SO(3,1)^{int}$ invariance follows the conservation of the six quantities

$$K_{\mu\nu} = (\Lambda S \Lambda^{-1})_{\mu\nu} = \Lambda_\mu{}^\rho S_{\rho\lambda} \Lambda_\nu{}^\lambda . \qquad (A.23)$$

The PB between $K_{\mu\nu}$ and any one of $J_{\mu\nu}$, P_μ vanishes, since the corresponding groups commute; among the $K_{\mu\nu}$ we have (compare with (A.18)!):

$$\{K_{\mu\nu}, K_{\alpha\beta}\} = g_{\nu\alpha}K_{\mu\beta} - g_{\mu\alpha}K_{\nu\beta} + g_{\nu\beta}K_{\alpha\mu} - g_{\mu\beta}K_{\alpha\nu} . \qquad (A.24)$$

There is an identity linking the Lagrangian and the Hamiltonian variables that is useful in checking some of the above statements. It results from the Poincaré invariance of L and reads:

$$u_\mu P_\nu - u_\nu P_\mu + S_\mu{}^\rho \sigma_{\rho\nu} - \sigma_\mu{}^\rho S_{\rho\nu} = 0. \qquad (A.25)$$

We stress that this is *not* a condition on the function f (ξ,η,ζ) in eq. (A.10) at all. In fact, if the expressions for P and S obtained from eq. (A.11) were to be substituted here, we would find that the resulting equation is identically obeyed for any f. In any case, this identity helps us see explicitly that $J_{\mu\nu}$ is conserved.

The development up to this point has used only the Poincaré and $SO(3,1)^{int.}$ invariances of L, and has expressed them in Hamiltonian language. We now pay attention to conditions (iii) and (iv) on L, and to the constraints of the theory. With any singular Lagrangian, the primary constraints are those relations among the q's and p's that arise on eliminating the velocities $\dot{q}$ from the equations defining the p's. In the H-R model, eqs. (A.11) show that both P_μ and $S_{\mu\nu}$ are exclusively functions of u_μ and $\sigma_{\mu\nu}$ alone. Every primary constraint must result on eliminating u and σ from these equations, and so must be a condition involving P and S alone; neither x nor Λ can occur. From the general theory of singular Lagrangian systems we know that chronometric invariance must lead to an explicitly Poincaré, and $SO(3,1)^{int.}$ invariant primary constraint $\Phi \approx 0$. Both P_μ and $S_{\mu\nu}$ are $SO(3,1)^{int.}$ invariants (cf. eq. (A.19b)); and the Lorentz invariants we can form from them are P^2, $\overline{V}^2$, W^2 and $S_{\mu\nu}S^{*\mu\nu}$. In view of eq. (A.21) we may equally well consider P^2, $\overline{V}^2$, $S_{\mu\nu}S^{*\mu\nu}$ and $S^2 = S_{\mu\nu}S^{\mu\nu}$

as the independent available Lorentz scalars out of which Φ must be formed. The situation is now similar to that in Chapter 6. If we express these quantities in terms of $f(\xi,\eta,\zeta)$ and its first partial derivatives, we have four quantities appearing as functions of only three variables. On elimination we expect to find (in the generic case!) one relation among P^2, $\overline{V}^2$, $S_{\mu\nu}S^{*\mu\nu}$ and S^2 which we may present in the form

$$\Phi = P^2 + \alpha(\tfrac{1}{2}S^2, \overline{V}^2, S_{\mu\nu}S^{*\mu\nu}) \approx 0. \tag{A.26}$$

This is the primary constraint that must exist for any Lagrangian obeying conditions (i), (ii) and (iii).

The condition (iv) on the Lagrangian in the H-R model can now be stated: it is that the four-vector $\overline{V}_\mu$ defined in eq. (A.22) must vanish as a primary constraint:

$$\overline{V}_\mu \approx 0. \tag{A.27}$$

If one computes P_μ and $S_{\mu\nu}$ in terms of $f(\xi,\eta,\zeta)$ and then evaluates $\overline{V}_\mu$, two groups of terms appear, respectively proportional to $\sigma_{\mu\nu}u^\nu$ and $\overset{*}{\sigma}_{\mu\nu}u^\nu$. Since one is dealing here with primary constraints, which is a stage prior to use of the Euler-Lagrange equations of motion or any of their consequences, one can secure eq. (A.27) only by arranging that the coefficients of $\sigma_{\mu\nu}u^\nu$ and $\overset{*}{\sigma}_{\mu\nu}u^\nu$ in $\overline{V}_\mu$ separately vanish. These lead to two nonlinear (Quadratic) partial differential equations on f, which cannot be solved in any simple way. In terms of L itself these conditions, given by Hanson and Regge, are:

$$2\,\frac{\partial L}{\partial a_1}\frac{\partial L}{\partial a_2} = \frac{\partial L}{\partial a_3}\left(a_1\frac{\partial L}{\partial a_1} + a_2\frac{\partial L}{\partial a_2} + a_3\frac{\partial L}{\partial a_3} + a_4\frac{\partial L}{\partial a_4}\right),$$

$$2\,\frac{\partial L}{\partial a_2}\frac{\partial L}{\partial a_3} = \frac{\partial L}{\partial a_1}\frac{\partial L}{\partial a_4} \tag{A.28}$$

This situation is to be contrasted with the vector model of Chapter 6. Since we are not going to solve these equations, we do not bother to rewrite them as equations for f. At any rate, the H-R model is characteri zed , at this stage, by any Lagrangian of the form (A.10) which also obeys these two partial differential equations.

Even though for the sake of a manifestly relativistic appearance we have written a set of four equations in (A.27), there are actually only three algebraically independent primary constraints here, since by definition $\overline{V}_\mu$ is orthogonal to P_μ. As a result of (A.27), the form of the previously established constraint (A.26) greatly simplifies. For any antisymmetric $S_{\mu\nu}$ one has the result

$$S^*_{\mu\nu}S^{\mu\rho} = \frac{1}{4}\delta_\nu{}^\rho\, S^*_{\alpha\beta}S^{\alpha\beta} . \tag{A.29}$$

We see then that the inner product of W_μ and $\overline{V}_\mu$ is

$$W.\overline{V} = \frac{1}{4}\, P^2\, S^*_{\alpha\beta}S^{\alpha\beta} . \tag{A.30}$$

So it follows that

$$\overline{V}_\mu \approx 0 \Rightarrow S^*_{\alpha\beta}S^{\alpha\beta} \approx 0 \tag{A.31}$$

also as a primary constraint. Thus we may omit the arguments $\overline{V}^2$ and $S_{\mu\nu}S^{*\mu\nu}$ in the function α in eq. (A.26); and we see that we have four independent primary constraints which for a relativistic appearance we present in the form

$$\Phi = P^2 + \alpha(\tfrac{1}{2}\, S^2) \approx 0, \quad \overline{V}_\mu = S_{\mu\nu}P^\nu \approx 0. \tag{A.32}$$

By the same token, eq. (A.21) simplifies to

$$W^2 \approx -\frac{1}{2}\, P^2S^2 , \tag{A.33}$$

so that $\Phi \approx 0$ is recognized as a constraint on the two Casimir invariants of the Poincaré algebra, i.e., as a mass-spin relation.

We can now set up an initial Hamiltonian, get phase space equations of motion, and carry out the constraint analysis. Since L is chronometric invariant, we have for Hamiltonian an expression

$$H = v\Phi + v_\mu\overline{V}^\mu, \quad v_\mu P^\mu = 0, \tag{A.34}$$

with arbitrary coefficients v and v_μ. (There are just four independent "unknown velocites" here). This Hamiltonian generates the following equations of motion (the one for $\dot{\Lambda}$ being rewritten as an expression for σ):

$$\dot{x}_\mu \approx \{x_\mu, H\} \approx 2vP_\mu - S_{\mu\nu}v^\nu ,$$

$$\sigma_{\mu\nu} \approx (\Lambda^{-1}\{\Lambda, H\})_{\mu\nu} \approx 2v\alpha'(\tfrac{1}{2}S^2)\, S_{\mu\nu} + v_\mu P_\nu - v_\nu P_\mu; \quad \text{(a)}$$

$$\dot{P}_\mu \approx 0,$$

$$\dot{S}_{\mu\nu} \approx \{S_{\mu\nu}, H\} \approx (P_\mu S_{\lambda\nu} - P_\nu S_{\lambda\mu})v^\lambda \quad . \quad \text{(b)} \tag{A.35}$$

We recognize that eqs. (A.35a) represent the result of trying to invert eqs. (A.11) to express u, σ in terms of P, S: at this stage, four velocites have remained unknown, and are embodied in v, v_μ. Equations (A.35b) coincide in content with the Euler-Lagrange equations of motion (A.16). We now impose the consistency conditions that Φ and V_μ remain zero for all s. The former is automatically obeyed since Φ happens to be first class:

$$\{\Phi, \bar{V}_\mu\} \approx 0 \Rightarrow \dot{\Phi} \approx 0. \tag{A.36}$$

As for V_μ, we find from the $S_{\mu\nu}$ equation of motion, or equally well from

$$\{\bar{V}_\mu, \bar{V}_\nu\} \approx P^2 \, S_{\mu\nu} \ , \tag{A.37}$$

that

$$\dot{\bar{V}}_\mu \approx 0 \Rightarrow S_{\mu\nu} \, v^\nu \approx 0. \tag{A.38}$$

The most general choice for v_μ obeying both restrictions (A.34), (A.38) can be found as follows. The tensor $S_{\mu\nu}$ viewed as an antisymmetric matrix must have rank four, two or zero. It cannot be either four or zero, since in the physically permitted configurations $S_{\mu\nu}$ does annihilate P_μ, and $S_{\mu\nu} \neq 0$. Thus the rank must be two, implying the existence of two independent vectors annihilated by $S_{\mu\nu}$. One may be taken to be P_μ, and the other is seen to be W_μ since from eqs. (A.29, A.31) we have:

$$S_{\mu\nu} W^\nu = S_{\mu\nu} S^{*\nu\rho} P_\rho = -\frac{1}{4} S^*_{\alpha\beta} S^{\alpha\beta} P_\mu \approx 0. \tag{A.39}$$

v_μ is thus necessarily some linear combination of P_μ and W_μ; but since v_μ must be, and W is, orthogonal to P_μ, we come up with

$$v_\mu = \frac{4v'}{P^2} W_\mu \ , \tag{A.40}$$

with v' arbitrary. The constraint analysis terminates at this point: there are no secondary constraints, and of the four initially unknown velocities only two, embodied in v and v', remain free at the end. The final Hamiltonian, necessarily a combination of all primary first class constraints, is

$$H = v\Phi + v'S^{*}_{\mu\nu}S^{\mu\nu} \; . \tag{A.41}$$

Thus of the four primary constraints (A.32), two have turned out to be first class: these are Φ and the linear combination W.V of the V's. The remaining two algebraically independent components of V_μ, constituting the part of V_μ orthogonal to both P_μ and W_μ, must be necessarily second class. (This is also clear from the fact that $S_{\mu\nu}$ has rank two: see eq. (A.37)).

The final equations of motion generated by (A.41) are quite a bit simpler than (A.35). To get the σ equation in a neat form we must use the result

$$W_\mu P_\nu - W_\nu P_\mu \approx P^2 \; S^{*}_{\mu\nu} \; , \tag{A.42}$$

which is proved by applying the duality operation twice to the left hand side. Then the equations of motion are

$$\dot{x}_\mu \approx 2vP_\mu \; , \qquad \sigma_{\mu\nu} \approx 2v\alpha'(\tfrac{1}{2}S^2)S_{\mu\nu} + 4v'S^{*}_{\mu\nu} \; ,$$

$$\dot{P}_\mu \approx 0, \qquad \dot{S}_{\mu\nu} \approx 0. \tag{A.43}$$

Thus at the end we have gained more constants of motion in $S_{\mu\nu}$, and the "trajectory function" α appears only in the equation for $\dot{\Lambda}$. In this sense, the spinorial model, and even the vectorial one, lead to more interesting structures.

Let us count the effective number of phase space degrees of freedom at this point. The full phase space is of dimension twenty, so on imposing the four primary constraints we get a constraint hypersurface of dimension sixteen. We might now choose two gauge constraints to pin down the unknown velocities v and v'; one of them, such as for example $x^\circ - s \approx 0$, would be conjugate to the first class constraint Φ, and another one would be conjugate to W.V. This would leave us with fourteen degrees of freedom, still two more than the twelve expected from the nonrelativistic counting. There is thus a need for two more independent constraints: it is important to realize that their role is only to reduce the number of independent variables from fourteen to twelve, and not in any sense to serve as conjugates to the two primary second class constraints. Thus these two conditions are not gauge constraints but rather are what are called invariant relations--these are by definition constraints compatible with already existing equations of motion, and are not intended to help fix unknown velocities in the Hamiltonian. To recover the proper physical picture, we need then one gauge constraint to fix v, and three more constraints comprising two invariant relations and one gauge constraint to fix v'. From the physical point of view, the latter set of three conditions evidently reduces the number of independent parameters in Λ from six to three, since all the other constraints mentioned do the job of ensuring that out of x_μ, P_μ and $S_{\mu\nu}$, only $\vec{x}$, $\vec{P}$ and S_{jk} are independent.

It is convenient to first develop the complete solution to the equations of motion (A.43), and then explain the choice of additional constraints in the H-R model. Unlike the situation in the vectorial and spinorial models, the general solution must now involve two unknown functions of s. Initial values $x_\mu(0)$, P_μ, $\Lambda(0)^\mu{}_\nu$, $S_{\mu\nu}$ at $s = 0$ may be chosen in any way consistent with the vanishing of Φ and V_μ; this places restrictions only on P_μ and $S_{\mu\nu}$. Thereafter, only $x(s)$ and $\Lambda(s)$ have to be solved for. The solution for $x(s)$ is trivial:

$$x_\mu(s) = x_\mu(0) + \phi(s)\, P_\mu \ ,$$

$$\dot{\phi} = 2v, \qquad \phi(0) = 0. \tag{A.44}$$

The SO(3,1) matrix $\Lambda(s)$ evolves in such a way as to make $K = \Lambda S \Lambda^{-1}$ independent of s. So if we write

$$\Lambda(s) = \Lambda(0)\,\Lambda_1(s), \tag{A.45}$$

then $\Lambda_1(s)$ is an SO(3,1) matrix in the little group of the tensor $S_{\mu\nu}$:

$$\Lambda_1(s)\, S = S\Lambda_1(s). \tag{A.46}$$

Now since $S_{\mu\nu}$ and $S^*_{\mu\nu}$ are antisymmetric, we can view them as belonging to the SO(3,1) Lie algebra (in the four-vector representation), and by exponentiation build up finite SO(3,1) transformations. The proper notation to do this is expressed by

$$(\exp(aS))^\mu{}_\nu = \sum_{n=0}^{\infty} \frac{a^n}{n!} (S^n)^\mu{}_\nu \ ,$$

$$(S^n)^\mu{}_\nu = S^\mu{}_\rho\, (S^{n-1})^\rho{}_\nu \ , \tag{A.47}$$

and similarly with exp (bS^*). Thus for any real a, b, exp (aS) and exp (bS^*) are SO(3,1) matrices. Moreover, because of eq. (A.31), these elements of SO(3,1) commute with one another, so the set of elements exp (aS) exp (bS^*) defines an Abelian two-parameter subgroup of SO(3,1). It can now be verified that this subgroup is indeed the little group of $S_{\mu\nu}$ (provided of course that $S_{\mu\nu}$ does not vanish identically!), so $\Lambda_1(s)$ must be of the form

$$\Lambda_1(s) = \exp(a(s)S + b(s)S^*). \tag{A.48}$$

With the help of the following easily established matrix identities (which hold since $S^{*\alpha\beta}S^{\alpha\beta} \approx 0$)

$$S^3 = -\kappa^2 S, \quad S^{*2} \approx S^2 + \kappa^2.\mathbb{1} \; ,$$
$$\kappa^2 = \frac{1}{2}S_{\alpha\beta}S^{\alpha\beta} \; , \tag{A.49}$$

the exponential can be evaluated in finite terms:

$$[\exp(aS + bS^*)]^\mu{}_\nu = \mathrm{Cos\,h}\kappa b\; \delta^\mu{}_\nu + \frac{\mathrm{Sin}\;\kappa a}{\kappa} S^\mu{}_\nu + \frac{\mathrm{Sin\,h}\kappa b}{\kappa} S^{*\mu}{}_\nu$$
$$+ \frac{\mathrm{Cos\,h}\kappa b - \mathrm{Cos}\;\kappa a}{\kappa^2} S^\mu{}_\rho S^\rho{}_\nu \, . \tag{A.50}$$

If we put in the forms (A.45, A.48) into the equation of motion for Λ in (A.43), we get:

$$\sigma_{\mu\nu} = (\Lambda^{-1}\dot{\Lambda})_{\mu\nu} = (\Lambda_1^{-1}\dot{\Lambda}_1)_{\mu\nu} = \dot{a}S_{\mu\nu} + \dot{b}S^*_{\mu\nu} \approx$$
$$2v\alpha'(\tfrac{1}{2}S^2)S_{\mu\nu} + 4v'S^*_{\mu\nu} \, . \tag{A.51}$$

Thus with

$$a(s) = \alpha'(\kappa^2)\,\phi(s),$$
$$\dot{b}(s) = 4v', \quad b(0) = 0 \quad , \tag{A.52}$$

the evolution of Λ has been determined:

$$\Lambda(s) = \Lambda(0)\,\exp(\alpha'(\kappa^2)\,\phi(s)S)\,.\,\exp\,(b(s)S^*). \tag{A.53}$$

Thus $\Lambda(s)$ develops via two independent commuting one-parameter subgroups in SO(3,1), related by duality; the rate of growth of one factor is tied in with the motion of x while that of the other is totally independent (the b term driven by v').

We can now present and easily understand the combination of two invariant relations plus one gauge constraint conjugate to $S^{*}_{\mu\nu}S^{\mu\nu}$ introduced by Hanson and Regge. Define the Poincaré invariant $SO(3,1)^{int.}$ unit "time-like" four vector

$$\rho^{\mu} = \Lambda^{\mu}{}_{\nu} \frac{P^{\nu}}{\sqrt{-P^2}} \tag{A.54}$$

With the help of eqs. (A.50, A.53) we see that it evolves with respect to s as

$$\rho^{\mu}(s) = \rho^{\mu}(0) \cos h\, \kappa b(s) + \frac{\operatorname{Sin} h \kappa b(s)}{\kappa\sqrt{-P^2}} \Lambda(0)^{\mu}{}_{\nu} W^{\nu}. \tag{A.55}$$

Hanson and Regge adopt the following new constraint relations:

$$Y^{\mu} = \rho^{\mu} + g^{\mu o} \approx 0. \tag{A.56}$$

These comprise only three algebraically independent conditions since

$$(\rho_{\mu} + g_{\mu o})Y^{\mu} = 0 . \tag{A.57}$$

Notice that while these new constraints leave intact the manifest Poincaré invariance, the $SO(3,1)^{int.}$ invariance is lost at this point, and only an $SO(3)^{int.}$ invariance remains. Conditions (A.56) imposed on the already obtained general solution to the equations of motion can possibly have only two kinds of consequences: (i) some new conditions may arise on the initial values x(o), P, Λ(o), S which were hitherto restricted only by the vanishing of Φ and V_{μ}; (ii) conditions may be placed on the velocities v, v' or equivalently on a(s), b(s). We show that indeed what happens is that Λ(o) gets restricted, and both v' and b(s) must vanish.

At s = 0, condition (A.56) imposed on the definition (A.54) says

$$(\Lambda(o)P)^{\mu} = \sqrt{-P^2}\, g^{\mu}_{o} \tag{A.58}$$

If we recall the definition (6.7.15) of the pure Lorentz transformation Λ(P) that carries the rest frame four-vector $(\sqrt{-P^2}, 0,0,0)$ into P^{μ}, we see that Λ(o) is restricted to be of the form

$$\Lambda(o) = \Lambda(R(o))\Lambda(P)^{-1} , \tag{A.59}$$

where $\Lambda(R(o))$ is the SO(3,1) matrix corresponding to some spatial rotation R(o). In detail,

$$\Lambda(o)^{o}{}_{o} = P^{o}/\sqrt{-P^2} , \quad \Lambda(o)^{o}{}_{j} = -P_j/\sqrt{-P^2} ,$$

$$\Lambda(o)^{j}{}_{o} = -R(o)_{jk} P_k/\sqrt{-P^2} ,$$

$$\Lambda(o)^{j}{}_{k} = R(o)_{jk} + \frac{R(o)_{j\ell}P_{\ell}P_k}{\sqrt{-P^2}(P^{o}+\sqrt{-P^2})} . \tag{A.60}$$

The effective number of degrees of freedom, at this stage, is thirteen: four x_μ, three $\vec{P}$, three S_{jk}, and three parameters in R(o). The breaking of $SO(3,1)^{int.}$ invariance is evident from the restrictions on $\Lambda(o)$. The later adoption of a gauge constraint conjugate to ϕ, such as $x^{o} - s \underset{\sim}{\approx} 0$, reduces the number of degrees of freedom from thirteen to twelve, as desired on physical grounds. Assuming the form (A.59) for $\Lambda(o)$, $\rho(s)$ is

$$\rho^{\mu}(s) \underset{\sim}{\approx} g^{\mu}_{o} \cos h k\, b(s) + \frac{\sin h\, k b(s)}{k\sqrt{-P^2}}(\Lambda(R(o))\Lambda(P)^{-1}W)^{\mu} . \tag{A.61}$$

Now since P.W = 0, it follows that $\Lambda(P)^{-1}W$ has vanishing time component, so the time and space parts of ρ^{μ} are:

$$\rho^{o}(s) = \cos h\, kb(s),$$

$$\rho^{j}(s) = \frac{\sin h\, kb(s)}{k\sqrt{-P^2}} (\Lambda(R(o))\Lambda(P)^{-1}W)^{j} . \tag{A.62}$$

It is clear that (A.56) can be satisfied for all s only if b(s), and so v', vanishes. Thus these conditions have achieved just what was promised.

The complete set of equations of motion, constraints, and solution for the relativistic spherical top now appear as:

$$\dot{x}_{\mu} \approx 2vP_{\mu} , \quad \sigma_{\mu\nu} \approx 2v\alpha'(\tfrac{1}{2}S^2)\, S_{\mu\nu},$$

$$\dot{P}_{\mu} \approx 0, \quad \dot{S}_{\mu\nu} \underset{\sim}{\approx} 0 : \tag{a}$$

$$\Phi = P^2 + \alpha(\tfrac{1}{2}S^2) \approx 0, \qquad V_\mu \approx 0,$$

$$\rho_\mu + g_\mu{}^o \approx 0; \qquad \text{(b)}$$

$$x_\mu(s) \approx x_\mu(o) + \phi(s)P_\mu \quad ,$$

$$\Lambda(s) \approx \Lambda(R(s))\Lambda(P)^{-1} \quad ,$$

$$R(s) = R(o)\ \Lambda(P)^{-1}\ \exp(\alpha'(\tfrac{1}{2}S^2)\phi(s)S)\Lambda(P). \quad \text{(c)} \qquad \text{(A.63)}$$

We close our introduction to the work of Hanson and Regge with the following remarks: Hanson and Regge show how to pass to a system of (preliminary) DB's by eliminating the six independent second class constraints in the V_μ, P_μ conditions. Already this results in $\{x_\mu, x_\nu\}^*$ being nonzero, as in our vector model. The final DB's are obtained by adopting $x^o - s \approx 0$ and eliminating $\Phi \approx 0$. They then introduce "commuting" Newton-Wigner type coordinates and then quantize the free top. For the resulting system each integer spin s occurs (2s + 1) times. The questions of electromagnetic interactions and magnetic moment coupling are discussed by Hanson and Regge [HAN 1], [HAN 2], and carried forward in Hojman's thesis [HOJ 1].

APPENDIX B:

GALILEAN SUBDYNAMICS

The most successful theory of interactions, the "Feynman rules", describe collisions from an overall Minkowski space-time viewpoint. Dynamics in Minkowski space has been conventionally described as the change in time of a configuration given at one instant in time in a particular reference frame in Minkowski space. This Newtonian form of dynamics was called the instant form by Dirac [DIR 6]. At the same time, Dirac proposed two alternative forms of dynamics one of which he designated the *front form of dynamics.* Dirac proposed to consider a family of parallel tangent spaces to the light cone instead of the usual family of parallel spaces at various instants of time.

Superficially, for classical theory, it is not clear that the front form of dynamics is any better than the instant form of dynamics. However, the quantum version of the front form shows an important distinction from the classical case. This follows from the fact that of the three coordinates *in the front* x_+, x_1, x_2, (with x_- being the coordinate specifying the front)--the coordinate x_+, unlike x_1 and x_2, has the nature of a time, that is, the momentum conjugate to x_+, (denoted by P_-) has a spectrum confined to the open positive half-line. It is accordingly not permitted in the quantum version to assume a kinematics based upon specifying a point within the front. *A way out of this fundamental difficulty is afforded by the fact that there exists an eight-parameter subgroup of the Poincaré group which adjoins the operator P_- to the seven generators that classically leave the front invariant.* This subgroup has the group structure of nonrelativistic (Galilean) dynamics in two space-dimensions, together with a scaling operator. The momentum operators P_+ and P_- play the role of Galilean Hamiltonian and Galilean mass, respectively, and inherently possess the proper (positive) spectrum. We call this structure "Galilean subdynamics," or the "Galilean subworld." Unlike the instant subworld (associated with a six-parameter subgroup) the Galilean subworld contains dynamics within itself which properly fits into the Poincaré world. For the dynamics within this Galilean subworld one can use nonrelativistic quantum mechanics [BIE 2,3].

The front form has an appealing physical interpretation as the "infinite momentum frame"; this view, however, is not very precise, since the front form is by no means a contraction limit of the Poincaré group, yet some of the intuitive ideas are indeed correct, and more easily seen from this view. In effect, the Lorentz contraction shrinks the z-axis, so that system and its motions are confined to the transverse plane; the time dilation slows the motions down so that they become non-relativistic.

Regardless of such heuristic motivations, the fact is that the Poincaré group does contain an *eight* parameter sub-group isomorphic to the Galilei group G (in 2 spatial-and 1 time-dimensions), augmented by a mass operator P_+ and a scaling operator, M_{03}. Designating these eight generators collectively by F, where

$$F = \{P_i, P_+, P_-, J_{12} = J_3, K_{-i}; M_{03}\}, \quad (i = 1,2) \qquad \text{(B-1)}$$

one finds that these operators obey the commutation rules:

(a) (2 + 1) Galilei group generators: G

$[J_3, P_i] = i\varepsilon_{3ij} P_j,$

$[J_3, K_{-i}] = i\varepsilon_{ij} K_{-j},$

$[J_3, P_-] = 0$

$[K_{-i}, K_{-j}] = 0$

$[K_{-i}, P_j] = i\delta_{ij} P_+,$

$[K_{-i}, P_-] = 2iP_i,$

$[P_i, P_j] = 0,$

$[P_i, P_-] = 0$

(b) Mass generator: P_+ (B-2)

$[P_+, G] = 0$ (B-2b)

(c) Scaling generator: M_{03}

$[M_{03}, J_3] = 0,$

$[M_{03}, K_{-i}] = iK_{-i},$

$[M_{03}, P_i] = 0,$

$[M_{03}, P_\pm] = \pm iP_\pm.$ (B-2c)

These commutation rules can be recognized as the commutation relations of an extended Galilei group G (in two spatial dimensions) together with a dilation (scaling) operator M_{03}. For this interpretation one must identify the operator P_- as the Galilei group Hamiltonian (H_G) and the operator P_+ as the mass operator M_G for the Galilei group.*

---

* Note that the commutator $[K_{-i}, P_j]$ in eq. (B-2) introduces P_+, which *does not* leave the front ($x_-=0$) invariant. Classically the commutator becomes a Poisson bracket, for which the RHS *is* zero, as required. This important discrepancy is purely quantal, and results from the fact that P_+ and P_- both have only positive eigenvalues so that the front is *not well-defined* quantum mechanically.

Taking the Poincaré generators to be

$$P_\mu \equiv (\hbar/i)\,\frac{\partial}{\partial x_\mu}\,,$$

$$M_{\mu\nu} \equiv x_\mu P_\nu - x_\nu P_\mu + S_{\mu\nu}\,,$$

where $S_{\mu\nu}$ are generators of a Lorentz group relization based upon the two harmonic oscillators of the internal structure given by (2.1.17). We note that the generators $S_{\mu\nu}$ contained in (B.1),(B.2) are just those, (2.1.17), which are no more than linear in the conjugate variable π; that is they act like point transformations on ξ_1 and ξ_2.

We will seek solutions to this Galilean structure, which exploit the fact that these generators correspond to nonrelativistic quantum mechanics of interacting particles, in a two-dimensional plane. Corresponding to this interpretation we take the center-of-mass momenta, P_1 and P_2, to be sharp: $P_i \rightarrow p_i$. The operator P_+, in the Galilean plane corresponds to the total mass, which because of the scaling generator M_{03}, has continuous eigenvalues $p_+ > 0$. We will also take this generator to be sharp: $P_+ \rightarrow p_+ > 0$. The associated wave function is then:

$$\exp \frac{i}{h}\,(p_1 x_1 + p_2 x_2 - p_+ x_-)\,. \tag{B-3}$$

For the Galilean Hamiltonian we take the operator P_- to obey the subsidiary relation

$$P_- \psi = (i\hbar\, \partial/\partial\, x_+)\psi \tag{B-4}$$

and then specify P_- to have the form (using "aligned bosons", cf.(2.4.1)):

$$P_- \equiv H_G = H_{c.m.} + H_{internal}'$$

where

$$H_{c.m.} = \frac{P_1^{\,2} + P_2^{\,2}}{2P_+}$$

$$H_{int}(\mathcal{U}) = \frac{m_o^{\,2}}{2P_+}\,[a_1(\Lambda)\bar{a}_1(\Lambda) + a_2(\Lambda)\bar{a}_2(\Lambda) + 1] \tag{B-5}$$

and Λ is *any* Lorentz transformation which transforms $\mathcal{U}_o$ into $\mathcal{U}$. Note that $H_{int.}$ does not depend on which Λ one chooses to carry $\mathcal{U}_o$ into $\mathcal{U}$; that is, H_{int} is invariant under Lorentz transformations which leave

$\mathcal{U}$ invariant.

It is useful to express this internal Hamiltonian in terms of the original boson operators a_i. Using the explicit form, eq. (2.4.1) for $a_i(\Lambda)$ one finds (after some calculation) the form

$$H_{int} = (M_o{}^2/2P_+)\Big\{\mathcal{U}_+)[(\pi_1 - A_1)^2 + (\pi_2 - A_2)^2] + (1/\mathcal{U}_+)(\xi_1{}^2 + \xi_2{}^2)\Big\} \qquad \text{(B-6)}$$

where the A_i are defined to be

$$A_i = \frac{\mathcal{U}_2\xi_2 - \mathcal{U}_1\xi_1}{\mathcal{U}_+}\,,$$

$$A_2 = \frac{\mathcal{U}_1\xi_2 + \mathcal{U}_2\xi_i}{\mathcal{U}_+}\,. \qquad \text{(B-7)}$$

This form for the internal Hamiltonian has certain unusual, but essential, features which we wish to discuss. First, note that the velocity $\mathcal{U}_+$ acts as a scale factor in determining the "size" of π_i and ξ_i. The generator for this transformation is K_3 which obeys the commutation rules

$$[K_3, \pi_i] = +\frac{1}{2}\,\pi_i\,,$$

$$[K_3, \xi i] = -\frac{1}{2}\,\xi_i \qquad \text{(B-8)}$$

so that π_i and ξ_i scale oppositely under K_3 (as is necessary for the Heisenbefg commutator to be properly scale-invariant). Thus we see that the scaling transformation, for a finite boost denoted by $\mathcal{U}_+$, scales the internal coordinates:

$$\mathcal{U}_+ \text{ boost:}\quad \pi_i \to (\mathcal{U}_+)^{1/2}\,\pi_i, \qquad \text{(B-9)}$$

$$\xi_i \to (\mathcal{U}_+)^{-1/2}\,\xi_i\,. \qquad \text{(B-10)}$$

The second unusual feature of this Galilean Hamiltonian is that through the coupling via the "vector potential" A_i, the internal wave function is a function of the velocity $\mathcal{U}_1$ and $\mathcal{U}_2$, which one would expect to be parallel to the momentum p_1 and p_2 of the center-of-mass Galilean motion. Note, however, that the *eigenvalues* of the operator in brackets, $\{\ldots\}$ in Eq. (B-6) are nonetheless independent of the 4-velocity $\mathcal{U}$.

These unusual features of the Galilean Hamiltonian pose a problem as to the logic (and consistency) of the subdynamical approach. Clearly one can choose arbitrary eigenvalues for the three commuting momentum operators P_1, P_2, P_+; but these three data do *not* suffice to determine the 4-velocity u, for one still lacks the mass value (implied by P_-) which must be obtained from the Hamiltonian itself. Although the structure is indeed self-consistent, let us simplify matters by avoiding this direct procedure, and following an alternative path. Consider this same Galilean Hamiltonian to be given by

$$\begin{aligned} H_G &= H(\chi\theta\phi) \\ &= \frac{P_1^2 + P_2^2}{2P_+} + \frac{m_o^2}{2P_+}[a_1(\chi\theta\phi)\bar{a}_1(\chi\theta\phi) \\ &\qquad + a_2(\chi\theta\phi)\bar{a}_2(\chi\theta\phi) + 1], \end{aligned} \tag{B-11}$$

where $(\chi\theta\phi)$ now specifies a *fixed* 4-velocity $u = (\sinh\chi\cos\theta, \sinh\chi\sin\theta\cos\phi, \sinh\chi\sin\theta\sin\phi, \cosh\chi)$.

The Galilean world is to contain besides this Hamiltonian, $H(\chi\theta\phi)$, the seven additional generators:

$$\{P_+, P_1, P_2;\ L_{-1}, L_{-2}, L_{03};\ M_{12} = L_{12} + J_{12}(u)\}\ . \tag{B-12}$$

Note that only in the last generator, M_{12}, have we introduced an operator acting on the internal space, and this operator is an explicit function of $\chi\theta\phi$ that is, $J_{12} = J_{12}(u)$.

It is easily verified that these operators, together with $H(\chi\theta\phi)$, close on the commutation relations of the Galilei group representation for which the five operators: $P_+, P_1, P_2, H(\chi\theta\phi)$, and J_{12} (spin) have been brought to diagonal form. Denoting these eigenvalues by

$$(P_+, P_1, P_2, H = P_-) \to p, \tag{B-13}$$

$$J_{12} \to M,$$

we find for the wave function

$$\psi(p, JM; \chi\theta\phi) = e^{ip\cdot x} \langle \xi_1 \xi_2 \left| \frac{[a_1(\chi\theta\phi)]^{J+M}[a_2(\chi\theta\phi)]^{J-M}}{[(J+M)!\ (J-M)!]^{\frac{1}{2}}} \right| 0; \chi\theta\phi\rangle\ . \tag{B-14}$$

(Note that the label J is determined by the number of quanta: $N + 1 = 2J + 1$.)

From the Galilean Hamiltonian we find

$$H_G \to p_- = \frac{1}{2}\,[p_1{}^2 + p_2{}^2 + m_o{}^2(N+1)]; \tag{B-15}$$

hence, one obtains the relations

$$2p_+p - p_1{}^2 - p_2{}^2 \equiv p\cdot p \equiv m^2 = m_o{}^2(N+1). \tag{B-16}$$

Thus for arbitrary values of the parameters $(\chi\theta\phi)$, we have solutions defined in a Hilbert space labeled by these parameters. The momentum eigenvalue p is however *independent* of these parameters.

The problem is now to take this solution, from the Galilean world, into the Poincaré world. Does this solution, Eq. (B-14) belong to an irrep of P? Clearly the solution, Eq. (B-14), does possess the Poincaré invariant $p^2 \to m^2$. Consider then the second Poincaré invariant. We already know that the stability group for p is generated by the spin operators:

$$J_1' = \tfrac{1}{2}[a_1(p/m)\bar{a}_2(p/m) + a_2(p/m)\bar{a}_1(p/m)],$$

$$J_2' = \tfrac{1}{2}[a_1(p/m)\bar{a}_2(p/m) - a_2(p/m)\bar{a}_1(p/m)],$$

$$J_3' = \tfrac{1}{2}[a_1(p/m)\bar{a}_2(p/m) - a_2(p/m)\bar{a}_1(p/m)]. \tag{B-17}$$

We take the little-group generators to be J_i', but agree that to obtain a Poincaré irrep we choose the unit 4-vector $\mathcal{U}(\chi\theta\phi)$ to be identical to p/m. Thus we orient the two vectors p and u to be parallel, and agree that all Lorentz transformations are henceforth to be generated by $M_{\mu\nu} = L_{\mu\nu} + S_{\mu\nu}$. (Note that $S_{\mu\nu}$ is defined in a fixed frame, $(\chi\theta\phi) = 0$.)

It follows now that the little group is generated by J_i', and hence $W^2 \to m^2 j(j+1)$, where $J = \frac{1}{2}N$ (N being the number of quanta).

(This is the self-consistent solution to our original Galilean problem whose internal Hamiltonian depends parametrically through $\mathcal{U}$ on the 4-velocity (p/m) operator.)

Thus, by explicit construction, two harmonic oscillators belonging to G implement the construction of Regge bands for a composite object in P. The construction allows *only* time-like, positive-energy, non-zero mass Poincaré irreps.

We emphasize that it is no great feat to obtain any desired sequence of Poincaré irreps with mass arbitrarily related to spin; one can do this by fiat as one wishes. The point is that by such a construction--viewed in the Poincaré world--one learns no structural information whatsoever. Thus there is nothing remarkable, Poincaré-wise, in obtaining the sequence: $m^2 = m_o{}^2(2j+1)$. What is remarkable in the construction discussed above is that this whole set of Poincaré irreps, viewed in the Galilean subworld, form a coherent set of states generated from a single (Galilean) Hamiltonian. (The fact that this is possible, for all masses and spins, is a strict result that the condition, $p/m = u$, is *independent of mass.)*

The method of Galilean subdynamics is suggestive. By identifying the Galilean plane with the plane realizing the various geometries of the Freudenthal magic square, for example, one might hope to find interesting new possibilities for quantal structures [BIE 4].

(It is worth noting that in the review by Kogut and Susskind [KOG 1], for example, only seven of the eight generators of G are identified. All eight are essential for the construction given above.)

REFERENCES

AND 1 J. L. Anderson and P. G. Bergmann, Phys. Rev. 3, 965(1951); a recent survey see also.

BAC 1 H. Bacry, in *Lecons sur la theorie des groupes et les symmetries des particules elementaires* (Gordon and Breach, New York, 1966), p. 309.

BAR 1 V. Bargmann, L. Michel, V. L. Telegdi, Phys. Rev. Lett. 2(1959) 435.

BEC 1 R. Becker, Gött. Nachr. p. 39 (1945).

BIE 1 L. C. Biedenharn and J. D. Louck, *Angular Momentum in Quantum Physics*, Addison-Wesley, to be published.

BIE 2 L. C. Biedenharn, M. Y. Han, and H. van Dam, Phys. Rev. D 8, 1735(1973).

BIE 3 L. C. Biedenharn and H. van Dam, Phys. Rev. D 9, 471(1974).

BIE 4 L. C. Biedenharn and H. van Dam, p. 5 in *Proceedings of International Symposium on Mathematical Physics*, Mexico City, 1976.

BOH 1 A. Böhm, Phys. Rev. 145, 1212(1966); 175, 1767(1968); D3, 367; ibid 377 (1971).

BOH 2 A. Böhm, R. Teese, A. Garcia, J. S. Nilsson, Phys. Rev. D 15, 689(1977).

BOH 3 A. Böhm, in *Lectures in Theoretical Physics*, edited by A. O. Barut and W. E. Brittin (Gordon and Breach, New York, 1968), Vol. X-B.

COR 1 H. C. Corben, *Classical and Quantum Theories of Spinning Particles*, Holden-Day, San Francisco, 1968.

DIR 1 P. A. M. Dirac, Proc. R. Soc. London A322, 435(1971).

DIR 2 P. A. M. Dirac, Proc. R. Soc. London A328, 1(1972).

DIR 3 P. A. M. Dirac, p. 260 in *The Principles of Quantum Mechanics* Oxford University Press, Oxford, 1948.

DIR 4 P. A. M. Dirac, Ann. Inst. H. Poincaré 11, 15(1949).

DIR 5 P. A. M. Dirac, J. Math. Phys. 4, 901(1963).

DIR 6 P. A. M. Dirac, Rev. Mod. Phys. 21, 392(1949).

DIR 7 P. A. M. Dirac, Can. J. Math. 2, 129(1950); Proc. R. Soc. London A246, 326(1958); *Lectures on Quantum Mechanics*, Yeshiva University-Belfer Graduate School of Science (Academic, New York, 1964).

DIX 1 W. G. Dixon, Nuovo Cin. 38 (1965) 1616; 34 (1964) 317.

DRE 1 W. Drechsler and M. E. Mayer, *Gauge Theory of Strong and Electromagnetic Interactions Formulated on a Fiber Bundle of Cartan Type* (Lecture Notes in Physics, Springer, Berlin, 1977, Vol. 67)

DYS 1 For References see e.g., G. J. Dyson "Symmetry Groups in Nuclear and Particle Physics", Benjamin, N.Y. 1966.

EGU 1 T. Eguchi, Phys. Rev. Lett. 44, 126(1980).

FOL 1 L. L. Foldy and S. A. Wouthuysen, Phys. Rev. 78, 29(1950).

FOL 2 L. L. Foldy, Phys. Rev. 102, 568(1956).

FOR 1 L. H. Ford and H. van Dam, Nucl. Phys. B169, 126(1980).

FRA 1 P. H. Frampton, *Dual Resonance Models*, (Benjamin, New York, 1974).

FRD 1 D. M. Fradkin, Am. J. Phys. 34, 314(1966).

FRE 1 J. Frenkel, Z. Phys. 37 (1926) 243.

GEL 1 M. Gell-Mann, Phys. Lett., 8, 214 (1964).

GOO 1 R. H. Good, Phys. Rev. 125 (1962) 2112.

GUR 1 F. Gürsey in *Seminar on High Energy Physics and Elementary Particles* I.A.E.C., Vienna, Austria, 1965.

HAA 1 R. Haag, J. T. Lopuszanski and M. Sohnius, Nucl. Phys. B88, 257 (1975).

HAG 1 C. R. Hagen and W. J. Hurley, Phys. Rev. Lett. 24, 1282 (1970).

HAN 1 A. Hanson, T. Regge, and C. Teitelboim, Acad. Naz. Lincei 22, 2(1976).

HAN 2 A. J. Hanson and T. Regge. Ann. Phys. (N.Y.) 87, 498 (1974).

HOJ 1 A. J. Hojman, thesis Princeton University, 1975.

JOR 1 P. Jordan, Zeitschrift Ph. 94, 531(1935).

KOG 1 J. Kogut and L. Susskind, Phys. Reports 8C, 75 (1973).

KOM 1 A. Komar, Phys. Rev. D18, 1881,1887,3617 (1978).

MAC 1 W. D. MacGlinn, Phys. Rev. Lett. 12, 243(1964).

MAJ 1 E. Majorana, Nuovo Cimento 9, 335 (1932).

MAN 1 S. Mandelstam, Phys. Rep. 13C, 260(1964).

MEL 1 H. J. Melosh, Phys. Rev. D9, 1095 (1974).

MIC 1 L. Michel, Lectures on Polarization at CERN, 1975 unpublished.

MIS 1 C. W. Misner, K. S. Thorne, J. A. Wheeler *Gravitation*, p.426, W. H. Freeman, San Francisco, 1973.

MOT 1 N. F. Mott and H. S. W. Massey, *The Theory of Atomic Collisions*, Oxford U.P. 1965, p. 214.

MUK 1 N. Mukunda, E. C. G. Sudarshan, C. C. Chiang, to be published.

MUK 2 N. Mukunda, Ann. Phys. (N.Y.) 99, 408(1976); Physica Scripta 21, 783(1980).

MUK 3 N. Mukunda, H. van Dam, L. C. Biedenharn, Phys. Rev. D22, 1938 (1980).

NAM 1 Y. Nambu, Progr. Theor. Phys. Suppls. 37, 38, 368 (1966).

NEW 1 T. D. Newton, E. P. Wigner, Rev. Mod. Phys. 21, 400 (1948).

ORA 1 L. O'Raifeartaigh, Phys. Rev. 139(1965) B1052.

PEI 1 R. Peierls, *Surprises in Theoretical Physics*, Princeton University Press (Princeton, N.J. 1979).

PRY 1 M. H. L. Pryce, Proc. R. Soc. A195, 62(1948).

ROH 1 F. Rohrlich, Phys. Rev. D23, (1981).

SAK 1 A. D. Sakharov, Doklady Akad. Nauk S.S.S.R. 177,70 (1967).

SCH 1 E. Schrödinger, Sitzungsb. d. Berlin Akad., 418(1930).

SCW 1 J. Schwinger, p. 222 in *Quantum Theory of Angular Momentum*, L. C. Biedenharn, H. van Dam, Academic Press, N.Y.N.Y. 1965.

SCW 2 J. Schwinger, Ann. Physics 119, 192 (1979).

SEG 1 I. Segal, J. Functional Analysis, 1, 1 (1967).

SHI 1 Iu. M. Shirokov, Dohl. Akad. Nauk SSSR 94, 857 (1954), 97, 737(1954).

STA 1 L. P. Staunton and S. Browne, Phys. Rev. D 12, 1026 (1975).

STA 2 L. P. Staunton, Phys. Rev. D. 13, 3269(1976).

STA 3 L. P. Staunton, Phys. Rev. D. 10, 1760 (1974).

SUD 1 E. C. G. Sudarshan, K. T. Mahantappa, Phys. Rev. Lett. 14, 163; 458 (1958).

SUD 2 E. C. G. Sudarshan and N. Mukunda, *Classical Mechanics in a Modern Perspective*, Wiley, New York, New York., 1978.

SUT 1 L. G. Suttorp and S. R. deGroot, Nuovo Cim. 65 245(1970).

TAK 1 T. Takabayasi, Suppl. Prog. Theor. Phys. 41, 130 (1968).

TAK 2 T. Takabayasi, Suppl. Prog. Theor. Phys. 59, 2133(1978).

TAK 3 T. Takabayasi, Suppl. Prog. Theor. Phys. 67, 1(1979).

THO 1 L. H. Thomas, Phys. Rev. 85, 868(1952).

THO 2 B. Bakamjian and L. H. Thomas, Phys. Rev. 92, 1300 (1953).

THO 3 L. H. Thomas, in *Quantum Theory of Atoms, Molecules and the Solid State*, edited by Per-Olov Löwdin (Academic, New York, 1966), cf. pp. 93-96.

TOD 1 I. T. Todorov, Comm. JINR E2-10125, Dubna (1976).

VAN 1 H. van Dam and Th. W. Ruijgrok, Physica 104A, 281 (1980).

VAN 2 H. van Dam and L. C. Biedenharn, Phys. Lett. 62B, 190 (1976).

VAN 3 H. van Dam and L. C. Biedenharn, Phys. Rev. C 14, 405 (1976).

VAN 4 H. van Dam and L. C. Biedenharn, p. 155 in *Group Theoretical Methods in Physics*, eds. Beiglböch, A. Bohm, E. Tahasuhi, (Springer, New York, N.Y., 1979).

VAN 5 H. van Dam and L. C. Biedenharn, Phys. Lett. 81B, 313 (1979).

VAN 6 H. van Dam and H. Wospakrik, to be published.

VAN 7 H. van Dam and M. Veltman, Nucl. Phys. B22, 397 (1970); G.R.G. 3, 215(1972).

VAN 8 H. van Dam, L. C. Biedenharn, and N. Mukunda, Phys. Rev. D23, 1451 (1981).

YUK 1 H. Yukawa, Phys. Rev. 77, 219(1950), ibid. 80, 1047(1950).

WER 1 J. Werle, ICTP Preprint, Trieste, 1965, unpublished.

WES 1 J. Wess, B. Zumino, Nucl. Phys. B70, 39(1970).

WEY 1 J. Weyssenhoff and A. Raabe, Acta Phys. Polon. 9, (1947) 7.

WIG 1 E. P. Wigner, Ann. Math. 40, 149 (1939).

Lecture Notes in Physics

Vol. 136: The Role of Coherent Structures in Modelling Turbulence and Mixing. Proceedings 1980. Edited by J. Jimenez. XIII, 393 pages. 1981.

Vol. 137: From Collective States to Quarks in Nuclei. Edited by H. Arenhövel and A. M. Saruis. VII, 414 pages. 1981.

Vol. 138: The Many-Body Problem. Proceedings 1980. Edited by R. Guardiola and J. Ros. V, 374 pages. 1981.

Vol. 139: H. D. Doebner, Differential Geometric Methods in Mathematical Physics. Proceedings 1981. VII, 329 pages. 1981.

Vol. 140: P. Kramer, M. Saraceno, Geometry of the Time-Dependent Variational Principle in Quantum Mechanics. IV, 98 pages. 1981.

Vol. 141: Seventh International Conference on Numerical Methods in Fluid Dynamics. Proceedings. Edited by W. C. Reynolds and R. W. MacCormack. VIII, 485 pages. 1981.

Vol. 142: Recent Progress in Many-Body Theories. Proceedings. Edited by J. G. Zabolitzky, M. de Llano, M. Fortes and J. W. Clark. VIII, 479 pages. 1981.

Vol. 143: Present Status and Aims of Quantum Electrodynamics. Proceedings, 1980. Edited by G. Gräff, E. Klempt and G. Werth. VI, 302 pages. 1981.

Vol. 144: Topics in Nuclear Physics I. A Comprehensive Review of Recent Developments. Edited by T.T.S. Kuo and S.S.M. Wong. XX, 567 pages. 1981.

Vol. 145: Topics in Nuclear Physics II. A Comprehensive Review of Recent Developments. Proceedings 1980/81. Edited by T. T. S. Kuo and S. S. M. Wong. VIII, 571-1.082 pages. 1981.

Vol. 146: B. J. West, On the Simpler Aspects of Nonlinear Fluctuating. Deep Gravity Waves. VI, 341 pages. 1981.

Vol. 147: J. Messer, Temperature Dependent Thomas-Fermi Theory. IX, 131 pages. 1981.

Vol. 148: Advances in Fluid Mechanics. Proceedings, 1980. Edited by E. Krause. VII, 361 pages. 1981.

Vol. 149: Disordered Systems and Localization. Proceedings, 1981. Edited by C. Castellani, C. Castro, and L. Peliti. XII, 308 pages. 1981.

Vol. 150: N. Straumann, Allgemeine Relativitätstheorie und relativistische Astrophysik. VII, 418 Seiten. 1981.

Vol. 151: Integrable Quantum Field Theory. Proceedings, 1981. Edited by J. Hietarinta and C. Montonen. V, 251 pages. 1982.

Vol. 152: Physics of Narrow Gap Semiconductors. Proceedings, 1981. Edited by E. Gornik, H. Heinrich and L. Palmetshofer. XIII, 485 pages. 1982.

Vol. 153: Mathematical Problems in Theoretical Physics. Proceedings, 1981. Edited by R. Schrader, R. Seiler, and D.A. Uhlenbrock. XII, 429 pages. 1982.

Vol. 154: Macroscopic Properties of Disordered Media. Proceedings, 1981. Edited by R. Burridge, S. Childress, and G. Papanicolaou. VII, 307 pages. 1982.

Vol. 155: Quantum Optics. Proceedings, 1981. Edited by C.A. Engelbrecht. VIII, 329 pages. 1982.

Vol. 156: Resonances in Heavy Ion Reactions. Proceedings, 1981. Edited by K.A. Eberhard. XII, 448 pages. 1982.

Vol. 157: P. Niyogi, Integral Equation Method in Transonic Flow. XI, 189 pages. 1982.

Vol. 158: Dynamics of Nuclear Fission and Related Collective Phenomena. Proceedings, 1981. Edited by P. David, T. Mayer-Kuckuk, and A. van der Woude. X, 462 pages. 1982.

Vol. 159: E. Seiler, Gauge Theories as a Problem of Constructive Quantum Field Theory and Statistical Mechanics. V, 192 pages. 1982.

Vol. 160: Unified Theories of Elementary Particles. Critical Assessment and Prospects. Proceedings, 1981. Edited by P. Breitenlohner and H.P. Dürr. VI, 217 pages. 1982.

Vol. 161: Interacting Bosons in Nuclei. Proceedings, 1981. Edited by J.S. Dehesa, J.M.G. Gomez, and J. Ros. V, 209 pages. 1982.

Vol. 162: Relativistic Action at a Distance: Classical and Quantum Aspects. Proceedings, 1981. Edited by J. Llosa. X, 263 pages. 1982.

Vol. 163: J. S. Darrozes, C. Francois, Mécanique des Fluides Incompressibles. XIX, 459 pages. 1982.

Vol. 164: Stability of Thermodynamic Systems. Proceedings, 1981. Edited by J. Casas-Vázquez and G. Lebon. VII, 321 pages. 1982.

Vol. 165: N. Mukunda, H. van Dam, L.C. Biedenharn, Relativistic Models of Extended Hadrons Obeying a Mass-Spin Trajectory Constraint. Edited by A. Böhm and J.D. Dollard. VI, 163 pages. 1982.